G-Proteins

G-Proteins
Signal Transduction and Disease

edited by

Graeme Milligan

Departments of Biochemistry and Pharmacology,

and

Michael Wakelam

Department of Biochemistry,
University of Glasgow, Glasgow, Scotland UK

ACADEMIC PRESS
Harcourt Brace Jovanovich, Publishers

LONDON   SAN DIEGO   NEW YORK   BOSTON
SYDNEY   TOKYO   TORONTO

This book is printed on acid-free paper

ACADEMIC PRESS LIMITED
24–28 Oval Road
LONDON NW1 7DX

United States Edition published by
ACADEMIC PRESS INC.
San Diego, CA 92101

A catalogue record for this book is available from the British Library

ISBN 0–12–497515–1

Typeset by Bath Typesetting Limited, Bath
Printed in Great Britain by The University Press, Cambridge

Contents

CHAPTER THREE

GTP-binding Proteins and Cardiovascular Disease
K. Urasawa and P.A. Insel

CHAPTER FOUR

**Alterations in G-protein-mediated Cell Signalling in Diabetes
Mellitus**
C.J. Lynch and J.H. Exton

CHAPTER FIVE

Adrenal Dysfunction and G-protein-mediated pathways
J.R. Hadcock and C.C. Malbon

CHAPTER SIX

G-proteins in Obesity
N. Bégin-Heick and N. McFarlane-Anderson

CHAPTER SEVEN

Thyroid Disorders
D. Saggerson

CHAPTER EIGHT

Alcoholism: a Possible G-protein Disorder
A.S. Gordon, D. Mochly-Rosen and I. Diamond

CHAPTER NINE

**The Role of G-proteins in the Regulation of Myeloid Haemo-
poietic Cell Proliferation and Transformation**
P. Musk, C.M. Heyworth, S.J. Vallance and A.D. Whetton

Contributors

Nicole Bégin-Heick, Department of Biochemistry, University of Ottawa, Ottawa, K1H 8M5, Canada.

Ivan Diamond, Departments of Pharmacology and Paediatrics, University of California, San Francisco, California, USA.

John H. Exton, Department of Molecular Physiology and Biophysics, Vanderbilt University Medical School, Nashville, TN 37232 USA.

Adrienne S. Gordon, The Ernest Gallo Clinic and Research Center, San Francisco General Hospital, San Francisco, CA 94110 USA.

John R. Hadcock, Diabetes and Metabolic Diseases Research Program, Department of Pharmacology, School of Medicine, Health Sciences Center, SUNY, Stony Brook, New York 11794-8651 USA.

Present address: Molecular and Cellular Biology Group, Agricultural Research Division, American Cyanamid Company, PO Box 400, Princeton, New Jersey 68543-0400 USA

Clare M. Heyworth, Experimental Haematology Department, Paterson Institute, Christie Hospital and Holt Radium Institute, Manchester, M20 9BX, UK.

Paul A. Insel, Department of Pharmacology, University of California at San Diego, La Jolla, San Diego, CA 92093 USA.

Michael A. Levine, Johns Hopkins University School of Medicine, Ross Research Building, Room 1029, 725 North Wolfe Street, Baltimore, MD 21205 USA.

Christopher J. Lynch, Department of Cellular and Molecular Physiology, The Milton S. Hershey Medical Center, The Pennsylvania State University College of Medicine, Hershey, PA 17033 USA.

Craig C. Malbon, Diabetic and Metabolic Disease Research, Department of Pharmacology, School of Medicine, Health Sciences Center, SUNY, Stony Brook, NY 11794-8651 USA.

Norma McFarlane-Anderson, Department of Biochemistry, University of Ottawa, Ottawa, K1H 8M5, Canada.

Graeme Milligan, Molecular Pharmacology Group, Departments of Biochemistry and Pharmacology, University of Glasgow, Glasgow, G12 8QQ, UK.

Alexander Miric, Johns Hopkins University School of Medicine, Ross Research Building, Room 1029, 725 North Wolfe Street, Baltimore, MD 21205 USA.

Daria Mochly-Rosen, Department of Neurology, University of California, San Francisco, California, USA.

Philip Musk, Department of Biochemistry and Applied Molecular Biology, U.M.I.S.T., Sackville Street, Manchester, M60 1QD, UK.

David Saggerson, Departments of Biochemistry and Molecular Biology, University College London, Gower Street, London, WC1E 6BT.

Kazushi Urasawa, Department of Pharmacology, University of California at San Diego, La Jolla, CA 92093 USA.

Susan J. Vallance, Department of Biochemistry and Applied Molecular Biology, U.M.I.S.T., Sackville Street, Manchester, M60 1QD, UK.

Anthony D. Whetton, Department of Biochemistry and Applied Molecular Biology, U.M.I.S.T., Sackville Street, Manchester, M60 1QD, UK.

Preface

Cellular signalling processes which require the involvement of guanine nucleotide binding proteins (G-proteins) have been widely studied. Basic research in this area has elucidated a considerable base of knowledge on the structure, diversity and function of these proteins and on the interaction of these proteins with the other components of transmembrane signalling cascades.

With this experimental framework in place considerable interest and emphasis has been placed, in recent years, on the assessment of potential alterations of hormone and neurotransmitter function in disease states. This book is designed to provide the reader with reviews of the results of such studies in a wide range of pathophysiological states. As will become clear, mutations in the genes encoding G-protein polypeptides have been observed in a number of systems and are likely to represent the primary lesion leading to the development of the clinical condition. However, the role of genetics in many other disease states are either controversial, undefined or subject to complex control and alterations in G-protein-linked signalling cascades which probably result as a consequence of rather than defining the development of the disease. Even if this is the case it is no less important to study these alterations as a clear understanding of these effects may well indicate useful strategies for therapeutic intervention. It is also clear that the bulk of studies on G-protein-mediated signalling and disease to date have focussed largely on alterations in function of the adenylyl cyclase cascade. This is largely a reflection that assays for the presence and function of the G-proteins involved in the control of cyclic AMP production have been widely available and relatively easy to perform.

With the recent identification of families of G-proteins responsible for receptor regulation of classes of phospholipases and indeed of G-proteins whose function remain undefined, it is clear that potential alterations in the level of expression and function of these proteins will soon be the target for studies in clinical biochemistry. Such studies are likely to progress rapidly and lead to a clearer understanding of abnormalities of G-protein-mediated signalling in disease. It is hoped that the chapters in this book may introduce the clinical investigator to this potentially important area and give other readers an indication of some of the medical relevance of this field of research.

CHAPTER ONE

Guanine Nucleotide-binding Proteins in Health and Disease: an Overview

GRAEME MILLIGAN
Molecular Pharmacology Group, Departments of Bio-
chemistry and Pharmacology, University of Glasgow,
Glasgow G12 8QQ, Scotland, UK

1 PREAMBLE

Guanine nucleotide-binding proteins of both the classical hetero-trimeric type and of the class of single chain polypeptides control a myriad of events in cells (Bourne *et al.*, 1990, 1991). They have been identified in species ranging from prokaryotes to man and the high degree of conservation of sequence of individual or at least homologous G-proteins throughout species attests to evolutionary pressure to maintain particular features of these proteins to regulate specific functions. Given the central position of the heterotrimeric G-proteins in the transduction of information it is not surprising that cells have developed means to regulate the function and expression of these proteins as one mechanism within a framework of approaches to modulate the effectiveness of signal transduction.

The overall theme of this book is to describe and discuss disease states in which alterations in the effectiveness of cellular signalling processes which are controlled by G-proteins have been observed. In many of the cases it is not yet clear whether these alterations represent primary effects which lead to the observed clinical manifestations or whether they occur subsequently as a result of, and in response to, other perturbations of cellular function. The purpose of this chapter will be to provide the reader with some basic information as to the variety and structure and function of the known G-proteins and to indicate how signalling processes in normal states can be altered by control of the activity and cellular levels of these polypeptides.

G-Proteins
ISBN 0-12-497515-1

2 CHARACTERISTICS OF G-PROTEIN-REGULATED TRANSMEMBRANE SIGNALLING CASCADES

Early experiments which demonstrated hormonal regulation of adenylyl cyclase failed to note that the presence of GTP was integral to this process. ATP used as the substrate for the adenylyl cyclase reaction was contaminated with GTP to a sufficient extent that further addition of this nucleotide was not required. The availability of more highly purified samples of ATP indicated, however, that the presence of GTP was indeed a prerequisite (see Birnbaumer (1990) for an historical overview). Such an observation indicated that a proteinaceous component which bound GTP with high affinity would be a central component of, at least, hormonal stimulation of cAMP production. Identification of this component was promoted by two separate strands of research. One of these related directly to our understanding of disease processes. A toxin isolated from *Vibrio cholerae*, which is the causative agent of the diarrhoeal disease cholera, was shown to elevate intracellular cAMP levels markedly, and in concert with this process a specific polypeptide which has been reported to be between 42 and 52 kDa incorporated [^{32}P]ADP-ribose if cholera toxin and cellular membranes were provided with [^{32}P]NAD$^+$ (Gill and Meren, 1978). The diarrhoea produced by this organism is due to efflux of Cl$^-$ and accompanying water from intestinal epithelial cells which is produced in response to elevated levels of intracellular cAMP. As covalent modification of this 45-kDa polypeptide caused a sustained activation of adenylyl cyclase, it seemed likely that this polypeptide would be the GTP-dependent coupling factor. The second line of research was the analysis of mutants of the S49 lymphoma cell line. Sustained elevated levels of cAMP are toxic to this cell line and thus selection of clones which, following treatment with mutagenizing agents, are able to survive treatment with stimuli known to elevate intracellular cAMP should provide cells deficient in one or other aspect of the hormonal activation of adenylyl cyclase. One clone isolated by this approach is now called cyc$^-$ (Bourne *et al.*, 1975) and when membranes of these cells were challenged with activated cholera toxin and [^{32}P]NAD$^+$, radiolabel was not incorporated into the putative GTP-binding protein which was identified in membranes of wild-type S49 cells. It has been demonstrated subsequently that the cyc$^-$ cells express neither the polypeptide nor indeed mRNA corresponding to this protein (Harris *et al.*, 1985) and on such a basis it is likely that the mutation is in the promoter elements which control the expression of the gene (see

Chapter 2 for a discussion of mutations of this gene which have been demonstrated to occur in at least certain patients with Albright's hereditary osteodystrophy).

Membranes of this cell line thus provided a sensitive reconstitution system for detection of the exogenous GTP regulatory protein and purification, initially from rabbit liver, of this protein was achieved (Northup *et al.*, 1980).

Early experiments on hormone binding to receptors and its regulation by GTP produced a potentially paradoxical situation in that GTP reduced the binding of these hormones even though it was required for the activation of the system (Rodbell *et al.*, 1971). Such observations, which are seen for agonists but not antagonists (Maguire *et al.*, 1976) at all G-protein-linked receptors, are a reflection that agonist affinity for receptor interaction is reduced by the presence of GTP and analogues of this nucleotide. The molecular rationale for the effect is that the complex of receptor and G-protein binds the hormone more tightly than the isolated receptor. GTP and its analogues dissociate the receptor–G-protein complex. Such differences in affinity may be as much as 1000-fold and provide a mechanism to empty the receptor of bound hormone following activation of the transduction cascade. Observations of reduced affinity of agonist binding in the presence of GTP has often provided the initial evidence for a role of a G-protein in the action of particular receptors prior to available information on the second messenger pathway regulated by the receptor.

Whilst early analysis of the hormonal regulation of adenylyl cyclase concentrated on the stimulation of this enzyme it was also clear that certain hormones could produce inhibition of cAMP production. Based partially on the type of theoretical considerations noted above, Rodbell (1980) proposed the likely existence of a separate G-protein which would mediate inhibitory regulation of adenylyl cyclase. Direct identification of this protein was again provided initially by consideration of the mode of action of a toxin associated with a disease process. A toxin isolated from *Bordetella pertussis*, the causative agent of whooping cough, was shown to modulate the secretion of insulin from pancreatic islet cells and as such was named islet activating protein (Katada and Ui, 1979). As the secretion of insulin is controlled by regulation of cAMP levels in the islet, it was suggested that islet activating protein (nowadays more commonly called pertussis toxin) might interfere with the generation of cAMP, and it was indeed noted that treatment of a variety of cells with the toxin led to an elevation of adenylyl cyclase activity (Katada and Ui, 1982). Parallel with this, membranes of cells treated with the toxin and [^{32}P]NAD$^+$ incorporated radioactivity into what appeared to be a single polypeptide of some 40 kDa (Katada and Ui, 1982). As pertussis toxin treatment attenuates receptor regulation of the inhibition of adenylyl cyclase, the 40-kDa polypeptide was assumed to be the inhibitory G-protein of the adenylyl cyclase cascade (G$_i$). Purification of this polypeptide was achieved either by following incorporation of [^{32}P]ADP-ribose from [^{32}P]NAD$^+$ catalysed by pertussis toxin (Bokoch *et al.*, 1983; Milligan and Klee, 1985) or the binding of [^{35}S]GTPγS (Sternweis and Robishaw, 1984) throughout purification schemes. However, in contrast to the situation for G$_s$α, no simple and appropriate functional reconstitution assay was available to follow the purification of 'G$_i$'.

3 THE STRUCTURE OF G-PROTEINS

During the initial purifications of both G$_s$ and 'G$_i$' it was apparent that a polypeptide of some 35 kDa co-purified was each of the ADP-ribosyl-transferase substrates. This polypeptide was thus demonstrated to be an integral part of the G-protein structure and named β subunit. Close examination of the region of gels close to the dye front subsequently showed the presence of a further polypeptide in all G-protein preparations which was in the region of 8–10 kDa. This polypeptide was named γ subunit. Although not covalently linked, β and γ subunits form a tightly associated complex which does not dissociate under other than denaturing conditions and as such the G-protein can be viewed functionally as a dimeric structure comprising α and βγ subunits. This heterotrimeric structure of α, β and γ subunits is shared between the G-proteins which transduce information between agonist-occupied receptors and second-messenger-generating enzymes. A further family of single-polypeptide GTP-binding proteins of between 21 and 30 kDa are widely expressed and function in a variety of both defined and undefined roles, including regulation of secretion and intracellular trafficking (Hall, 1990a,b). Whilst direct roles for at least the *ras* family of single-polypeptide G-protein in transmission of information from the extracellular milieu has been proposed (e.g. Wakelam *et al.*, 1986), this remains a contentious issue. None of the subunits of G-proteins which have been identified to date are likely to be trans-membrane polypeptides and as such the physical location of the G-proteins is at the cytoplasmic face of the plasma membrane. It was believed initially that the more hydrophobic nature of the βγ complex would act as the membrane anchor for the entire G-protein complex. Recent work has clarified mechanisms of membrane attachment to a greater degree. Analysis of cDNA clones has demonstrated that the predicted C-terminal sequence of G-protein γ subunits conforms to a so-called CAAX (cysteine–aliphatic–aliphatic–X) motif. Such sequences are known to be susceptible to complex post-transitional modification (Maltese, 1990) and analysis of such modifications has been examined in particular detail for the P21^{ras} proteins, as site-directed mutagenesis of this region alters the ability of these proteins to interact with and remain at the plasma membrane and also restricts or eliminates their oncogenic activity (Hancock *et al.*, 1989). In the P21^{ras} proteins, a C15 farnesyl group is added to the cysteine residue of the CAAX motif (Hancock *et al.*, 1989; Casey *et al.*, 1989), the three terminal amino acids are removed and the new C-terminal cysteine residue becomes carboxymethylated. In the case of the G-protein γ subunits, processing of the C-terminal CAAX motif is probably produced in a similar manner (Backlund *et al.*, 1990) except that the cysteine residue becomes modified by the addition of a C20 geranyl-geranyl group (Mumby *et al.*, 1990; Yamane *et al.*, 1990) derived from intermediates of cholesterol biosynthesis. Whilst inhibitors of 3-hydroxy-3-methylglutaryl CoA reductase, which is the rate-limiting enzyme for cholesterol biosynthesis, would probably be too broad spectrum in action, there is interest in inhibitors of the specific isoprenyl transferase enzymes as a means to limit the attachment of CAAX-motif-containing proteins to various cellular membranes and hence potentially regulate signalling and growth control.

It is not only G-protein γ subunits which become modified by fatty acylation. Information from the isolation of cDNA clones corresponding to G-protein α subunits predicts that the majority, but not all, have the N-terminal sequence (Met)-Gly-X-X-X-Ser. Such a sequence represents the consensus sequence for N-myristoyltransferase (Grand, 1989). A number of G-proteins which contain this sequence have indeed been demonstrated directly to contain myristate and site-directed mutagenesis to alter the N-terminal glycine to alanine prevents myristoylation and limits membrane attachment (Jones *et al.*, 1990). The presence of serine at position 5 has been demonstrated to be essential for high-affinity interaction of defined peptides with the *N*-myristoyltransferase activity (Grand, 1989). In this regard it is worth noting that the α subunit of G$_s$ is not predicted to have a serine in this position and myristoylation of this polypeptide has not been demon-strated. Gα11, a recently identified polypeptide which has been shown to have the characteristics of a G-protein able to transduce infor-mation from agonist-occupied receptors to phosphoinositidase C (Smrcka *et al.*, 1991), has a markedly different predicted N-terminal sequence and as such is unlikely to be fatty acylated in this fashion.

It is generally assumed that receptor-mediated exchange of GTP for GDP in the nucleotide-binding site of a G-protein leads to the physical dissociation of α from $\beta\gamma$ subunits. Whilst much of the evidence for this model is derived from studies performed on purified G-proteins in detergent solution which have been activated by poorly hydrolysed analogues of GTP, there is little direct experimental evidence to dispute the concept that dissociation also occurs in the cell. Sites of interaction of α and $\beta\gamma$ subunits have not been defined clearly but the N-terminal regions of G-protein α subunits have been implicated in $\beta\gamma$ interactions (Neer and Clapham, 1988). Tryptic cleavage of the extreme N-terminal regions of a number of G-proteins prevents both interaction of α and $\beta\gamma$ subunits and the interaction of the α subunit with the membrane (Eide *et al.*, 1987). It cannot be concluded from such an analysis that interaction with the $\beta\gamma$ complex is essential for membrane interaction because proteolytic removal of the N-terminus will also remove the site of attachment of myristic acid and, as noted above, this modification is essential for membrane interaction. It is still an open question why activated G-proteins which are not modified by myristoylation, such as G$_s$, largely remain in contact with the membrane following activation. It has been noted that prolonged incubation with analogues of GTP will cause a release of G$_s\alpha$ from the membrane of cells (Milligan and Unson, 1989) but such observations require the maintenance of highly artificial conditions. It may be that other as yet undetected fatty acylations or other covalent modifi-cations serve to anchor such G-proteins to the cytoplasmic face of the plasma membrane.

4 GENETIC DIVERSITY OF G-PROTEINS

Subsequent to the identification of the G-proteins which regulate the activity of adenylyl cyclase and knowledge of the structure and function of transducin in rod outer segments a considerable number of further G-protein α subunits have been identified by either the purifi-

cation of the polypeptides or by isolation of cDNA clones corresponding to such polypeptides. To date some 16 distinct α subunits are
known and the observation that differential splicing mechanisms can
produce multiple forms of $G_o\alpha$ (Hsu *et al.*, 1990) indicates the potential
for further variation to be recorded. A number of the identified G-
proteins are expressed in very restricted tissue locations; for example,
G_{olf} appears to be present only in olfactory neuroepithelium (Jones and
Reed, 1989) and transducin 1 and transducin 2 in rod and cone outer
segments respectively (Lerea *et al.*, 1986). As such the likely functions
of such proteins are determined by their location, G_{olf} being responsible for communication of signals from odorants to adenylyl cyclase
and the forms of transducin in the transmission of information from
photon reception to the regulation of cGMP levels. By contrast, other
G-proteins such as G_i2, G_i3 and G_s are expressed in an essentially
universal manner in mammalian tissues and this is likely to mirror
their function. For example, G_s and G_i2 respectively regulate stimulation and inhibition of adenylyl cyclase. The functions of a number of
the G-proteins, such as G_i1, have been less clearly defined. However,
the high degree of sequence conservation of these polypeptides
between species suggests that they will be shown to have highly
specific roles. Furthermore, as G_i1 is expressed to detectable levels in
only a relatively small number of tissues, indications as to function
may be gained from consideration of distribution.

The identified number of distinct G-proteins has recently been
increased substantially by the use of polymerase chain reaction
technology. Partial sequences of G-protein α chains named G10–G14
(Strathmann *et al.*, 1989) were isolated by this strategy and this
information has allowed the isolation of full-length cDNA clones
corresponding to some of these polypeptides (Strathmann and Simon,
1990). As these G-proteins are not predicted to have the cysteine
residue four amino acids from the C-terminus which form the acceptor
site for ADP-ribose in pertussis toxin-sensitive G-proteins, these will
not be modified by this toxin and as such must be likely to participate
in signalling functions which are not disrupted by treatment with the
toxin. As the majority of systems in which hormonal regulation of
the activity of phospholipases has been recorded are insensitive to
pertussis toxin treatment, these novel G-proteins are candidates to fulfil
these functions. They are also widely expressed. The first pertussis and
cholera toxin-insensitive G-protein to be defined by cDNA cloning (G_z)
(Fong *et al.*, 1988; Matsuoka *et al.*, 1988) was also considered as a
potential candidate for a regulation of phosphoinositidase C. How-

ever, the restricted distribution of this gene product makes it unlikely that it would act in this manner and, as noted above, an examination of the functions of cells which do express this G-protein may provide useful hints as to its function.

Whilst the bulk of attention and experimental effort has been directed towards expanding the family of α subunits and defining their function, it should not be overlooked that genetic diversity also occurs at the level of both β and γ subunits. To date four distinct cDNA clones corresponding to β subunits have been isolated (e.g. Levine et al., 1990) and a combination of both molecular and immunological information indicates a similar number of γ subunits (Gautam et al., 1989, 1990; Robishaw et al., 1989). As it has been more difficult in general to define specific roles for the βγ complex in cellular signalling processes, it is perhaps not surprising that little information is currently available on whether distinct groupings of β + γ subunits perform individual functions. The bulk of experiments attempting to approach this question have simply isolated βγ complex associated with transducin, which at least at the level of β subunit appears to consist entirely or largely of one molecular species, and compared its function against βγ complex(es) isolated from brain, which is likely to be a mixture of individual forms, in reconstitutive assays. Differences in the activities of such preparations have been recorded but the functional significance of these observations is not yet clear. However, as certain functions, including the regulation of phospholipase A_2 activity in rod outer segments, seems clearly to be regulated by the availability of βγ rather than α subunits (Jelsema and Axelrod, 1987), such experiments are likely to be pursued with diligence. Given the potential for complex metabolism of arachidonic acid released from phospholipids by the action of phospholipase A_2 to a range of bioactive agents, the availability of βγ subunits to regulate this process could be expected to have a range of further consequences. The mechanism of muscarinic acetylcholine receptor regulation of K^+ channels in atrium is the most studied example. Debate has centred on the relative importance of α and βγ subunits in this process and it has been argued recently that the observed role of βγ subunits in activation of K^+ channels may well be a reflection of prior βγ activation of phospholipase A_2, which by causing the generation of 5-lipoxygenase metabolites from the arachidonic acid so produced leads to an activation of the channel (Kim et al., 1989; Kurachi et al., 1989). This story is unlikely to be concluded soon, however, as it has been reported recently that, rather than activating, βγ complex actually inhibits flux through this channel (Okabe et al., 1990).

The other major area in which a role has been proposed for the βγ complex is in the regulatory inhibition of adenylyl cyclase. The isolation of the α subunits of both G_s and 'G_i' associated with apparently identical βγ complexes promoted the concept that inhibitory regulation of adenylyl cyclase could occur by activation of an inhibitory receptor causing the dissociation of 'G_i' αβγ to provide a sink of βγ complexes which could then physically interact with and deactivate free activated G_sα and/or prevent the activation of further G_sα whilst activated G_sα decayed based solely on the intrinsic K_{cat} of its GTPase activity (review: Gilman, 1987). In such a model the role of the α subunit of 'G_i' would be more passive than is usually accepted to be the case for other G-protein α subunits. For such a model to be supported a number of conditions would have to be met. Firstly, physical dissociation of the G-protein heterotrimeric complexes must actually occur following receptor activation. As noted above, whilst this is generally held to be self-evident it has proved difficult to provide conclusive evidence either for or against this idea in membranes or in a whole cell. Secondly, the relative ratios of G_s and G_i must be such that activation of 'G_i' would provide a pool of free βγ which was stoichiometrically high in comparison to levels of G_sα. This generally appears to be the case but recent immunological estimations of the levels of G_sα indicate that levels of this polypeptide may be higher than was originally believed based upon incorporation of [^{32}P]ADP-ribose catalysed by cholera toxin. Thirdly, availability of excess βγ subunit should limit the lifetime of GTP on G_sα to promote or maintain the inactivation of this polypeptide as a regulator of adenylyl cyclase. Addition of βγ complex, in something of a molar excess, to purified or *E. coli* expressed G_sα, reduces substantially the measured GTPase activity (Graziano and Gilman, 1989). This might at first sight suggest that the lifetime of GTP-ligated G_sα is actually prolonged by interaction with βγ subunits but it must be appreciated that the GTPase assay is not limited kinetically at the measured step, i.e. hydrolysis of GTP, but rather by the exchange of GTP for GDP, which is the rate-limiting step of the GTPase cycle (see Milligan (1988) for details). It is believed that the observed effects of the addition of βγ result largely if not entirely from a reduction in the rate of GDP dissociation from G_sα. Thus the observed reduction in G_sα GTPase produced by βγ may be produced simply by preventing replacement of GDP by GTP. Fourthly, genetic deletion of G_sα should prevent receptor-driven inhibition of adenylyl cyclase. As noted above, the cyc$^-$ variant of S49 lymphoma cells does not express G_sα but somatostatin-mediated inhibition of adenylyl cyclase does occur (Jakobs and Schultz, 1983), indicating that either the βγ complex or 'G$_i$α' must have a direct interaction with adenylyl cyclase. This observation would imply that βγ regulation of the activity of G$_s$α cannot account entirely for the inhibitory control of adenylyl cyclase. Whilst beyond the scope of both this chapter and this book, genetic analysis in yeast has provided convincing evidence for the importance of a direct role of βγ subunits in the control of mating factor response.

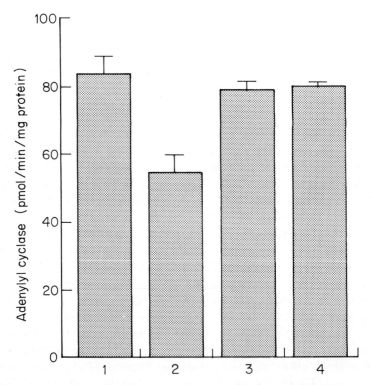

Figure 1. Hormonal inhibition of adenylyl cyclase is mediated via G$_i$2. Membranes from NG108-15 cells were incubated for 60 minutes with IgG fractions derived from either normal rabbit serum (1,2) or from an antiserum, AS7 (3,4), which in these cells identifies only G$_i$2α of the various G-proteins expressed. After this period forskolin (10 μM)-amplified adenylyl cyclase activity was assessed in the absence (1,3) or presence (2,4) of the opioid receptor agonist DADLE (10 μM). Results which are presented as mean ± SEM ($n = 4$) are adapted from McKenzie and Milligan (1990a) and demonstrate that receptor-mediated inhibition of adenylyl cyclase is abolished by antibodies which recognize the C-terminal decapeptide of G$_i$2α.

The use of antisera directed towards the extreme C-terminus of G-proteins has indicated that receptor contact with specifically G_i2 (presumably $\alpha\beta\gamma$) is required to produce inhibition of adenylyl cyclase (Simonds *et al.*, 1989; McKenzie and Milligan, 1990a) (Figure 1). Such experiments, however, cannot conveniently discriminate between whether it is the production of $G_i2\alpha$ or the $\beta\gamma$ complex which is responsible for inhibition of adenylyl cyclase but indicate only that activation of this heterotrimeric G-protein is both necessary and adequate to produce inhibition of adenylyl cyclase.

5 IDENTIFIED G-PROTEIN MUTATIONS: WHAT DO THEY TELL US ABOUT THE REGULATORY DOMAINS OF G-PROTEINS?

As noted above, the S49 cyc⁻ cell line lacks expression of $G_s\alpha$. Other defined mutations in this G-protein have been identified in clones derived from the wild-type S49 lymphoma cells. Of these, the mutation which provides the greatest insight into G-protein structure and function is the *unc* (uncoupled) mutation, in which the polypeptides corresponding to both β-adrenergic receptor and $G_s\alpha$ are present but are unable to physically interact with one another. Isolation and sequencing of a cDNA clone corresponding to $G_s\alpha$ from S49 *unc* cells demonstrated that there is only a single base difference between the wild-type and *unc* isoforms (Rall and Harris, 1987; Sullivan *et al.*, 1987). This alteration is predicted to cause the replacement of an arginine residue which is located six amino acids from the C-terminus of $G_s\alpha$ in the wild type with a proline residue in the mutant form. Given that the C-terminal region of G-protein α subunits, based on secondary structure predictions, is usually depicted as an amphipathic α helix, replacement of arginine by the helix breaker amino acid proline might be expected to produce a deleterious structural modification (Figure 2). This information coupled with the knowledge that pertussis toxin-catalysed ADP-ribosylation, which occurs on a cysteine residue four amino acids from the C-terminus, also attenuates contact between receptors and G-proteins which are substrates for this modification, demonstrated a central role for the C-terminal region in contacts between receptors and G-proteins. The substitution of a single amino acid in the α subunit of $G_s\alpha$ has also been recorded in the H21a mutant of S49 lymphoma cells (replacement of glycine 226 by alanine (G226A)).

This mutation produces a form of $G_s\alpha$ which is able to interact with receptor and bind guanine nucleotides but is unable to produce activation of adenylyl cyclase. This may be a reflection of the fact that the mutation prevents $G_s\alpha$ from adopting an active confirmation (Miller *et al.*, 1988). The conservative alteration of glycine to alanine might not, in general, be thought to be likely to result in a phenotypic modification but it is only one amino acid removed from a glutamine residue (Gln 227) which plays a central role in the GTPase activity of the G-protein (see below).

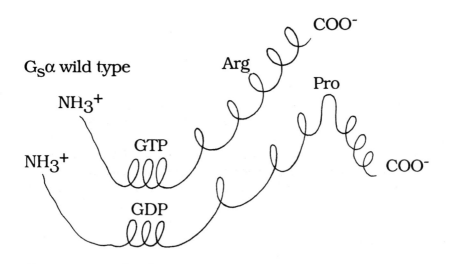

Figure 2. Diagrammatic representation of the structure of $G_s\alpha$ expressed in wild type and in the *unc* mutant of S49 lymphoma cells. A single base substitution which leads to the alteration of a single amino acid in $G_s\alpha$ appears to be sufficient to produce the *unc* mutation of S49 lymphoma cells (Sullivan *et al.*, 1987; Rall and Harris, 1987). The predicted alteration is the replacement of an arginine residue which is located six amino acids from the C-terminus of $G_s\alpha$ in the wild type by proline. Based on the helix breaker properties of proline this may provide sufficient distortion in the G-protein structure to prevent functional interactions of the mutated G-protein with receptors. Receptor-promoted exchange of GTP for GDP which is produced in the wild type will not occur in the *unc* mutant and G-protein activation of adenylyl cyclase will thus be prevented.

Table 1. Identified mutations in the α subunits of heterotrimeric G-proteins in human tumours

G-protein modified	Position and identity of alteration	Tissue	Reference
$G_s\alpha$	201 (Arg to Cys)	Pituitary	Landis *et al.* (1989)
$G_s\alpha$	201 (Arg to His)	Pituitary	Landis *et al.* (1989)
$G_s\alpha$	227 (Gln to Arg)	Pituitary	Landis *et al.* (1989)
$G_s\alpha$	227 (Gln to Leu)	Pituitary	Vallar (1990)
$G_s\alpha$	227 (Gln to His)	Thyroid	Lyons *et al.* (1990)
$G_i2\alpha$	179 (Arg to Cys)	Adrenal	Lyons *et al.* (1990)
	179 (Arg to His)		
$G_i2\alpha$	179 (Arg to Cys)	Ovary	Lyons *et al.* (1990)
	179 (Arg to His)		

Mutations in $G_s\alpha$ have also been recorded in man (Table 1). Vallar and co-workers (Vallar *et al.*, 1987) initially noted that a subset of patients with growth hormone-secreting pituitary adenomas appeared to have a constitutively activated adenylyl cyclase in these tumours. Because growth hormone-releasing hormone (GHRH) functions to cause release of growth hormone by stimulating production of cAMP, then the noted constitutive activation of adenylyl cyclase produced growth hormone release without a requirement for GHRH. Such a system would imply a loss of growth regulation in cells in which cAMP acts as a positive signal for cell proliferation. It was noticeable that tumours from these patients appeared similar in their regulation of adenylyl cyclase to that which would be expected in cells treated with cholera toxin. The locus of the alteration in these patients was defined to be $G_s\alpha$ as reconstitution of S49 cyc⁻ membranes with detergent extracts containing G_s from these patients was able to reproduce the constitutive activation of adenylyl cyclase observed in the native membranes. Based on the probability that the modification would represent a physical mutation in the $G_s\alpha$ polypeptide, Bourne named this modification *gsp* (based on the traditional nomenclature of onco-genes as three-letter descriptive identifiers) (Bourne, 1987) and sub-sequently Bourne, Vallar and collaborators demonstrated that such tumours did indeed harbour mutations (Landis *et al.*, 1989). The cloning and sequencing of $G_s\alpha$ cDNA from four tumours revealed mutations in all cases. Two distinct sites of mutation were identified. Either Arg 201 was replaced with either Cys or His (these mutations are thus designated R201C and R201H, based on the use of the single-letter code for amino acids) or Gln 227 was replaced by Arg (Q227R). Given the biochemical characteristics of the regulation of adenylyl cyclase in the tumours noted above it was perhaps not surprising that these two sites should be the modified ones. Arg 201 in $G_s\alpha$ is the site for ADP-ribosylation catalysed by cholera toxin and, as noted earlier, this modification produces a constitutive activation of adenylyl cyclase by inhibiting the GTPase turn-off mechanism. Gln 227 was also a likely site for an activating mutation which was based on inhibition of the GTPase activity of the G-protein. This site is positionally equivalent to Gln 61 of $p21^{ras}$ and it has been well established that replacement of this amino acid leads to a reduction in GTPase activity and a promotion of oncogenic activity. Furthermore it had also been noted that deliberate site-directed mutagenesis of $G_s\alpha$ at this position to any other amino acid allowed within the restrictions of the genetic code produces an inhibition of the GTPase activity.

As analysis of genomic DNA from the cases of pituitary adenoma which displayed constitutive activation of adenylyl cyclase demonstrated the presence of both wild-type and mutated sequences it was likely that the individuals had both a normal and a mutant $G_s\alpha$ allele and that this might be predicted to be a reflection of a somatic rather than a germ line mutation. Such a hypothesis was strengthened by the demonstration that white blood cells from two such patients had only the wild-type allele.

Transfection of $G_s\alpha$ Q227R into Swiss 3T3 fibroblasts activates DNA synthesis in such cells (Zachary et al., 1990), again showing the oncogenic potential of this mutation in cells in which cAMP acts as a positive growth regulator. (It is worth noting, however, that it is only in certain clones of Swiss 3T3 cells that cAMP does act as a positive regulator. In other clones of these cells it acts in an inhibitory manner.) It has also been recorded that transgenic mice which harbour the gene encoding the cholera toxin A1 ADP-ribosyltransferase polypeptide in pituitary somatotrophs demonstrate gigantism, elevated serum GH levels, somatotroph proliferation and pituitary hyperplasia (Burton et al., 1991).

It has also been noted recently that other G-protein α subunits can contain oncogenic mutations which may again result in a reduction in GTPase activity. Screening of a wide range of different human tumours for mutations in the $G_i2\alpha$ gene has shown substitutions at Arg 179 (which is at the equivalent position in this protein to Arg 201 in $G_s\alpha$) in tumours from adrenal cortex and ovarian granulosa tissue

(Lyons *et al.*, 1990). As inhibition of cAMP production appears to be a contributory mitogenic signal in a much wider range of tissues than is stimulation of cAMP production, and, as noted above, G_i2 appears to be the true inhibitory G-protein of the adenylyl cyclase cascade, this observation was not unexpected. It remains to be demonstrated but is highly probable that equivalent mutations will be detected in G-proteins which regulate the hydrolysis of phosphatidylinositol 4,5-bisphosphate and other lipid moieties which act as a reservoir of secondary signals as considerable evidence implicates strongly these pathways in the regulation of growth control.

6 HORMONAL REGULATION OF G-PROTEIN LEVELS

As noted in subsequent chapters, alterations in the circulating concentrations of either thyroid hormones (Chapter 7) (Figure 3) or a range of steroid hormones (Chapter 5) can produce considerable alterations in cellular steady-state concentrations of various G-proteins. Such observations have been recorded most regularly in defined animal models in which manipulation of hormone level can be accurately controlled or even ablated. Such alterations in G-protein levels have been correlated with alterations in either sensitivity to or maximal effectiveness of the activity of agents which function by the activation of 'seven transmembrane element receptors' which have been shown or are believed to be transduced from agonist binding to the effector system by one or other of the G-proteins which are modulated in amount. It has been more difficult to reproduce such effects within the human population, even in patient groups in which known alterations in hormonal sensitivity would appear to account for, or certainly to contribute to, the observed clinical manifestation. Such limitations may of course be related to the large genetic pool that such patient groupings are almost invariably drawn from but it does mean that animal and cell culture model systems remain the prime experimental material used to understand the underlying biochemical modifications. Hormonal effects on G-protein steady-state levels are not restricted to the action of the lipid-soluble agents noted above which would be expected to produce their effects by the regulation of gene expression. Indeed, a series of recent reports has examined hormonal regulation of G-protein levels in response to hormones which activate cell surface receptors rather than entering the cell directly. It is well established that agonist activation of such receptors usually leads to a down-regulation of the receptor polypeptide, but concurrent regulation of the cellular G-protein population has not been observed on a regular basis.

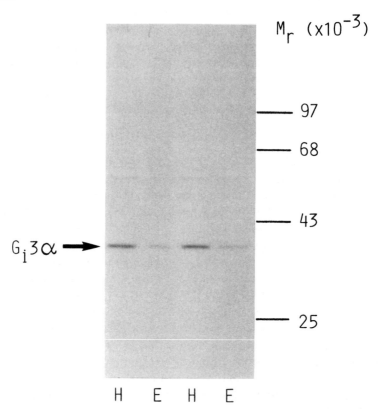

Figure 3. Elevated levels of $G_i3\alpha$ in membranes of white adipocytes from hypothyroid rats. Percoll purified plasma membranes from white adipocytes from hypothyroid (H) and euthyroid (E) rates were immunoblotted to detect the α subunit of G_i3. Membranes from the hypothyroid rats had two-fold higher levels of $G_i3\alpha$ than membranes from the euthyroid animals. Similar results were found for the α subunits of both G_i1 and G_i2 (see Milligan and Saggerson (1990) for details).

Lipolysis in adipocytes is regulated by levels of cAMP. Incubation of rat adipocytes in primary culture with phenylisopropyladenosine (PIA) (an adenosine deaminase-resistant analogue of adenosine), which is a potent anti-lipolytic agent because it mediates inhibition of adenylyl cyclase, produces a marked reduction in levels of the α subunits of each of the 'G$_i$-like' G-proteins (G$_i$1, G$_i$2 and G$_i$3) which are expressed by these cells. Indeed, sustained incubation with receptor-saturating levels of PIA reduces cellular levels of both G$_i$1α and G$_i$3α by some 90% and levels of both G$_i$2α and β subunit by some 50% (Green *et al.*, 1990). By contrast, levels of G$_s\alpha$ are not altered significantly. Removal of the agonist allows recovery of levels of each of these polypeptides, demonstrating that this downregulation is not a reflection of a metabolic run-down of the adipocytes in culture. This *in vitro* model does have a direct *in vivo* equivalent. Chronic treatment of rats with PIA delivered by a minipump system indicates a similar downregulation of 'G$_i$-like' G-proteins in adipocytes subsequently isolated from these animals compared to sham-treated controls (Longabaugh *et al.*, 1989). The mechanism of such G-protein downregulation remains to be firmly established, but as no obvious alteration in steady-state levels of mRNA encoding these polypeptides has been noted, it is probable that transcriptional control plays little or no part. At least within the primary culture model, time courses of downregulation of the adenosine A$_1$ receptor and the G-proteins are similar and it is possible that co-internalization and degradation of receptor and G-proteins occurs. Inhibition of adenylyl cyclase in adipocytes is also produced by prostanoid agonists of the E series and by nicotinic acid. Whilst treatment of adipocytes in primary culture with prostaglandin E$_1$ causes a similar downregulation of the 'G$_i$-like' G-proteins, treatment with concentrations of nicotinic acid which are equally effective in the inhibition of lipolysis has no detectable effect on cellular G-protein levels (Green *et al.*, 1991). Whilst it is unknown what the natural agonist is for the receptor activated by nicotinic acid, it is clear that it is a G-protein-linked receptor as nicotinic acid-mediated inhibition of adenylyl cyclase is both GTP-dependent and abolished by pretreatment of the cells with pertussis toxin.

A second system in which activation of a cell surface receptor leads to marked alteration in the cellular steady-state levels of a G-protein is neuroblastoma × glioma hybrid, NG108-15, cells. Activation of an IP-class receptor on the surface of these cells with either iloprost, which is a stable analogue of prostacyclin, or with prostaglandin E$_1$ leads to a substantial reduction in cellular levels of G$_s\alpha$ without altering levels of the α subunit of any of G$_i$2, G$_i$3 and G$_o$ (McKenzie and Milligan, 1990b). Time courses of the loss of the IP receptor and G$_s\alpha$ are identical (Figure 4), and whilst concentration curves for agonist-dependent loss of the G-protein are identical to the occupancy of the receptor and the

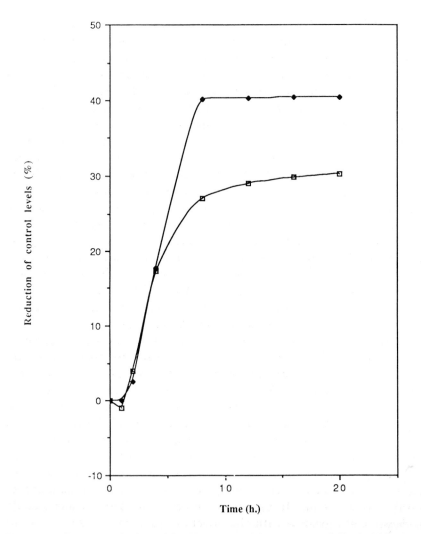

Figure 4. Downregulation of the IP prostanoid receptor and $G_s\alpha$ in NG108-15 cells in response to prostanoid agonists are temporally coincident. NG108-15 cells in tissue culture were treated with prostaglandin E_1 (10 μM) for varying times. Relative levels of $G_s\alpha$ (□) and the IP prostanoid receptor (◆) were then measured in membranes derived from the cells.

activation of adenylyl cyclase and hence of the generation of cAMP, the reduction in G-protein levels is independent of cAMP production. Again, loss of the G-protein does not involve a reduction in levels of $G_s\alpha$ mRNA and as effects of agonist and cycloheximide on cellular levels of $G_s\alpha$ are additive it is likely that reduction in either the rate of transcription and/or translation of pre-existing mRNA can contribute little to the observed phenomenon. It remains to be established clearly if the loss of $G_s\alpha$ in this system results from an enhanced rate of proteolytic degradation following activation of the receptor. However, this possibility is not without precedent. As noted above, cholera toxin catalyses the transfer of ADP-ribose to the α subunit of G_s and produces a persistent activation of the G-protein. However, due to the availability of specific antisera it has been established recently that cholera toxin treatment of cells produces a marked downregulation of levels of $G_s\alpha$ (Milligan *et al.*, 1989; Chang and Bourne, 1989). This appears to occur in every cell in which the phenomenon has been examined and in some systems is so severe that the expected permanence of the activation of adenylyl cyclase is not observed (Macleod and Milligan, 1990) (Figure 5), presumably due to the fact that the residual level of G_s is insufficient to activate fully the cellular complement of adenylyl cyclase. The available evidence suggests that this downregulation of $G_s\alpha$ is likely to result from enhanced proteolytic degradation but this remains to be established clearly.

In contrast to the examples cited above, there are examples in which agonist-mediated alterations in G-protein levels are produced, at least primarily, by transcriptional control. As discussed fully in Chapter 8, prolonged exposure of NG108-15 cells to moderately high concentrations of ethanol produces a loss of some 35% of cellular levels of $G_s\alpha$ and in concert there is a similar percentage loss of $G_s\alpha$ function and mRNA (Mochly-Rosen *et al.*, 1988). The effect of ethanol is believed to occur due to inhibition of the nucleoside transporter which causes an accumulation of extracellular adenosine which can then act as an agonist at the adenosine A_2 receptor to stimulate the production of cAMP (Nagy *et al.*, 1990). These observations are mechanistically somewhat at variance with the effects of prostanoid agonists noted above. Furthermore, it has also been recorded that direct activation of the adenosine A_2 receptor on NG108-15 cells with the agonist NECA does not produce a detectable downregulation of $G_s\alpha$ (McKenzie *et al.*, 1991).

The fact that prostanoid agonists, but not NECA, produce a substantial downregulation of $G_s\alpha$ in NG108-15 cells may appear paradoxical as the receptors for both agents act to stimulate adenylyl cyclase. The discrimination between these observations may relate to the relative levels of the two receptors. Whilst it has not been possible to assess levels of the adenosine A_2 receptor accurately, as agents like [3H]-

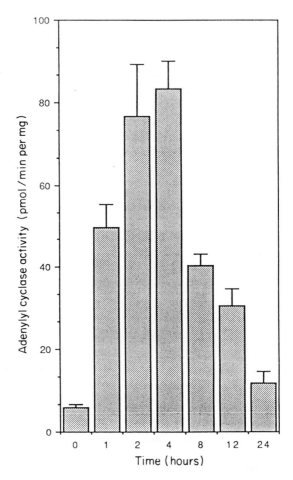

Figure 5. Biphasic regulation of adenylyl cyclase in NG108-15 cells by cholera toxin. NG108-15 cells in tissue culture were treated for varying times with cholera toxin (100 ng/ml), harvested and membranes prepared. Adenylyl cyclase activity in the absence of other regulators was then assessed. Results are presented as mean $\pm$ SD and are taken, with slight modification, from Macleod and Milligan (1990).

NECA bind to non-receptor sites as well as to the receptor, levels of the IP receptor are high. The specific binding of [^{3}H] prostaglandin E_1 to membranes of NG108-15 cells indicates that there are some 1.5 pmol receptor/mg membrane protein. Prolonged activation of the receptor with prostaglandin E_1 or iloprost produces downregulation of some 1.0 pmol receptor/mg membrane protein and in concert a loss of between 5 and 7 pmol $G_s\alpha$/mg membrane protein. If activation of the adenosine A_2 receptor was to produce downregulation of the receptor and $G_s\alpha$ with similar stoichiometry then the loss of G_s might represent only a very small percentage of the total and hence be experimentally impossible to detect.

Perhaps the most interesting example to date of hormonal regulation of steady-state levels of G-proteins has been reported following activation of adenylyl cyclase in the S49 lymphoma cell line. Whether stimulated indirectly by the activation of a β-adrenergic receptor or directly by the addition of forskolin, a 25% reduction in levels of $G_s\alpha$ was noted and, more interestingly, levels of $G_i2\alpha$ were increased three-fold. Levels of $G_i2\alpha$ mRNA were also elevated and the time course of elevation of $G_i2\alpha$ mRNA slightly preceded that of the polypeptide (Hadcock *et al.*, 1990). As similar results were not produced in a mutant of the S49 lymphoma cell line (kin⁻) which lacks cAMP-dependent protein kinase activity it could be concluded that the elevated levels of G_i2 resulted from a cAMP-dependent protein kinase-catalysed promotion of either transcription of the $G_i2\alpha$ gene or of stability of the relevant mRNA (Figure 6). However, as elevated levels of $G_i2\alpha$ are not observed in all cells when treated with forskolin, either this effect is not of universal significance or the steady-state levels of $G_i2\alpha$ are sufficiently low in S49 lymphoma cells that the stimulation of production of G_i2 provides a large percentage increase in this system but only a marginal percentage increase in other systems in which this question has been examined.

The functional consequences of hormonal regulation of steady-state levels of G-proteins also shed light on the relevance of hormonal regulation of G-protein levels. In concert with the upregulation of G_i2 in S49 cells in response to cAMP, somatostatin, which acts to inhibit adenylyl cyclase, is able to produce a greater inhibition of this activity (Hadcock *et al.*, 1990). As it is clear that activation of adenylyl cyclase can alter cellular levels of a number of receptors, such results must be treated with a degree of caution, but it does appear that elevated levels of G_i2 in S49 cells correlates with a greater functional response of receptors which utilize this G-protein and as such this provides further supporting evidence for the argument that G_i2 is the inhibitory G-protein of the adenylyl cyclase cascade. Such information cannot be usefully extracted from the use of anti-lipolytic hormones to cause the downregulation of G_i subtypes in adipocytes in primary culture as all three gene products are downregulated concurrently. However, the loss of these G-proteins following treatment with PIA and prostaglandin E_1 but not with nicotinic acid allows assessment of the role of hormone-mediated downregulation of G-proteins in the development of desensitization. Treatment with nicotinic acid causes desensitization of the anti-lipolytic response to subsequent challenge with nicotinic acid but not to either PIA or prostaglandin E_1. As such the desensitization produced by nicotinic acid is homologous in nature. By contrast, prior exposure to either of the two other anti-lipolytic agents results in desensitization to these two agents and also to nicotinic acid, and as such is a heterologous form of desensitization. The most obvious conclusion from these studies is that the loss of G-proteins is required for the development of heterologous but not homologous desensitization (Green *et al.*, 1991).

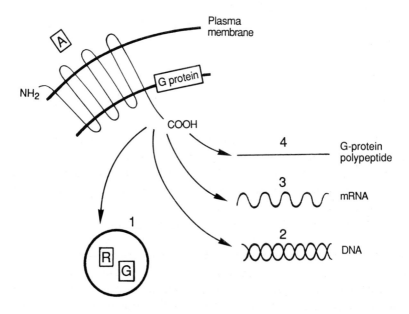

Figure 6. Mechanisms for the regulation of G-protein levels in response to receptor activation. In a range of cells, agonist activation of receptors can produce an alteration in steady-state levels of G-proteins. Mechanisms may include: (1) co-internalization of receptor and G-protein; (2) alterations in the rate of transcription of G-protein genes; (3) alteration in the stability of mRNA encoding G-proteins; (4) alterations in G-protein turnover (see Milligan and Green, 1991).

As noted in Chapter 3, control of cAMP levels regulates the contractility of the heart. β-Adrenergic stimulation is observed to promote

contraction and cAMP phosphodiesterase inhibitors are used thera-
peutically to maintain cAMP levels to treat congestive heart failure.
A range of studies has indicated potential alterations in G-protein
levels associated with the failing heart although it is not clear that an
obvious consensus has been reached from the studies to date (see
Chapter 3). As exposure of cardiac myocytes in tissue culture to
noradrenaline for 48 h has been reported to cause an upregulation of
pertussis toxin-sensitive 'G_i-like' G-proteins (Reithmann *et al.*, 1989)
which would act to reduce cellular levels of cAMP, it might appear
that maintenance of elevated cAMP levels using the type of treatments for congestive heart failure noted above could be expected to
produce as many potential hazards in the medium term as benefits in
the short term. Such conflicting indications highlight the concerns
about how to devise and predict the most appropriate therapeutic
intervention.

7 HORMONAL REGULATION OF G-PROTEIN ACTIVITY

Given the central role of G-proteins in trans-plasma membrane information transfer it would be surprising if cells had not developed
strategies based around reversible covalent modification to allow
alterations in the functional activity of these proteins. Phosphorylation catalysed by serine/threonine kinases and subsequent removal by
phosphatases could provide such control mechanisms. Indeed, phosphorylation of G_i2 in particular has been recorded in a number of
systems. Whilst phosphorylation of pertussis toxin-sensitive G-proteins by both the insulin receptor kinase (Krupinski *et al.*, 1988) and
the cAMP-dependent protein kinase (Watanabe *et al.*, 1988) has been
recorded in situations in which the purified kinase and purified G-
proteins are co-incubated, there may be little relevance to either of
these effects *in vivo*. It has not been possible to see phosphorylation of
'G_i' by the addition of insulin to cells, and whilst addition of insulin to
membranes of cells can result in a reduction in the amount of incorporation of [^{32}P]ADP-ribose into a 40-kDa polypeptide catalysed by pertussis toxin (Rothenberg and Kahn, 1988), the concentration of insulin
required to produce this effect is non-physiological and it has not been
demonstrated that the phenomenon is actually a reflection of activation of the insulin receptor. The noted effect of cAMP-dependent
protein kinase in causing phosphorylation of 'G_i' is also surprising

as there is no consensus sequence for the action of this kinase in the primary sequence of the 'G$_i$-like' G-proteins. More convincingly, relevant phosphorylation has been noted in cells including U937 cells (Issakani et al., 1989) and rat hepatocytes (Bushfield et al., 1990a) by the addition either of tumour-promoting phorbol esters or of Ca^{2+}-mobilizing hormones which are known to cause the production of sn 1–2 diacylglycerol, the endogenous activator of some of the isozymes of protein kinase C. The functional significance of such protein kinase C-catalysed phosphorylation is more difficult to assess. As noted in Chapter 4, G$_i$2 in hepatocytes isolated from rats which have been treated with streptozotocin to induce diabetes do not incorporate ^{32}Pi from $\gamma[^{32}P]$ATP into G$_i$2 in response to activation of protein kinase C whilst hepatocytes from untreated animals do (Bushfield et al., 1990b). This is a reflection that in the hepatocytes from the diabetic animal the G-protein is already in the phosphorylated state. In this state, low concentrations of Gpp(NH)p, a poorly hydrolysed analogue of GTP, are unable to produce inhibition of adenylyl cyclase whereas they do in membranes of hepatocytes from the control animals. Such results would appear to indicate that phosphorylation of G$_i$2 causes a functional inactivation of this protein but such conclusions are clouded by the observation that whilst both Gpp(NH)p- and GTP-mediated inhibition of adenylyl cyclase are attenuated in adipocytes from streptozotocin-treated diabetic rats, receptor-driven inhibition of adenylyl cyclase in response to any of PIA, E-series prostaglandins or nicotinic acid is paradoxically not abolished and may even be more effective than in adipocyte membranes from untreated animals (Strassheim et al., 1990).

The interrelationships of diabetes (Chapter 4), obesity (Chapter 6) and cardiovascular disease (Chapter 3) indicate the need to examine the phosphorylation state and functionality of G-proteins and the effectiveness of G-protein-linked signalling cascades in these disease states.

8 PERSPECTIVES

This chapter has attempted to provide the non-specialist with an overview on the structure and function of heterotrimeric G-proteins and to indicate how studies into the function and cellular regulation of these proteins in the normal state can provide insights into how alterations in the expression or function of these proteins which may occur in disease states can contribute to the pleiotropic effects seen in the cellular signalling machinery of the body. In many cases the biochemical observations are still largely phenomenological but it is clear that significant progress in the detailed analysis of these perturbations will be forthcoming and may contribute significantly to our understanding of the disease and to strategies designed to alleviate the symptoms.

ACKNOWLEDGEMENTS

Work in the author's laboratory is supported by the Medical Research Council, the Agriculture and Food Research Council and the British Heart Foundation.

REFERENCES

Backlund, P.S. Jr, Simonds, W.F. and Spiegel, A.M. (1990). *J. Biol. Chem.* **265**, 15572–15576.

Birnbaumer, L. (1990). *FASEB. J.* **4**, 3068–3078.

Bokoch, G.M., Katada, T., Northup, J.K., Hewlett, E.L. and Gilman, A.G. (1983). *J. Biol. Chem.* **258**, 2072–2075.

Bourne, H.R. (1987). *Nature* **330**, 517–518.

Bourne, H.R., Coffino, P. and Tomkins, G. (1975). *Science* **187**, 750–752.

Bourne, H.R., Sanders, D.A. and McCormick, F. (1990). *Nature* **348**, 125–132.

Bourne, H.R., Sanders, D.A. and McCormick, F. (1991). *Nature* **349**, 117–127.

Burton, F.H., Hasel, K.W., Bloom, F.E. and Sutcliffe, J.G. (1991). *Nature* **350**, 74–77.

Bushfield, M., Murphy, G.J., Lavan, B.E., Parker, P.J., Hruby, V.J., Milligan, G. and Houslay, M.D. (1990a). *Biochem. J.* **268**, 449–457.

Bushfield, M., Griffiths, S.L., Murphy, G.L., Pyne, N.J., Knowler, J.T., Milligan, G., Parker, P.J., Mollner, S. and Houslay, M.D. (1990b). *Biochem. J.* **271**, 365–372.

Casey, P.J., Solski, P.A., Der, C.J. and Buss, J.E. (1989). *Proc. Natl. Acad. Sci. USA* **86**, 8323–8327.

Chang, F.-H. and Bourne, H.R. (1989). *J. Biol. Chem.* **264**, 5352–5357.

Daniel-Issakani, S., Spiegel, A.M. and Strulovici, B. (1989). *J. Biol. Chem.* **264**, 20240–20247.

Eide, B., Gierschik, P., Milligan, G., Mullaney, I., Unson, C., Goldsmith, P. and Spiegel, A. (1987). *Biochem. Biophys. Res. Commun.* **148**, 1398–1405.

Fong, H.K.W., Yoshimoto, K.K., Eversole-Cire, P. and Simon, M.I. (1988). *Proc. Natl. Acad. Sci. USA* **85**, 3066–3070.

Gautam, N., Baetscher, M., Aebersold, R. and Simon, M.I. (1989). *Science* **244**, 971–974.

Gautam, N., Northup, J., Tamir, H. and Simon, M.I. (1990). *Proc. Natl. Acad. Sci. USA* **87**, 7973–7977.

Gill, D.M. and Meren, R. (1978). *Proc. Natl. Acad. Sci. USA* **75**, 3050–3054.

Gilman, A.G. (1987). *Ann. Rev. Biochem.* **56**, 615–649.

Grand, R.J.A. (1989). *Biochem. J.* **258**, 625–638.

Graziano, M. and Gilman, A.G. (1989). *J. Biol. Chem.* **264**, 409–418.

Green, A., Johnson, J.L. and Milligan, G. (1990). *J. Biol. Chem.* **265**, 5206–5210.

Green, A., Milligan, G. and Belt, S.E. (1991). *Biochem Soc. Trans.* **19**, 212S.

Hadcock, J.R., Ros, M., Watkins, D.C. and Malbon, C.C. (1990). *J. Biol. Chem.* **265**, 14784–14790.

Hall, A. (1990a). In: *G-proteins as Mediators of Cellular Signalling Processes* (Houslay, M.D. and Milligan, G., eds), pp. 173–195. Wiley and Sons, Chichester.

Hall, A. (1990b). *Science* **249**, 635–640.

Hancock, J.T., Magee, A.I., Childs, J.E. and Marshall, C.J. (1989). *Cell* **57**, 1167–1177.

Harris, B.A., Robishaw, J.D., Mumby, S.M. and Gilman, A.G. (1985). *Science* **229**, 1274–1277.

Hsu, W.H., Rudolph, U., Sanford, J., Bertrand, P., Olate, J., Nelson, C., Moss, L.G., Boyd, A.E., Codina, J. and Birnbaumer, L. (1990). *J. Biol. Chem.* **265**, 11220–11226.

Jakobs, K.H. and Schultz, G. (1983). *Proc. Natl. Acad. Sci. USA* **80**, 3899–3902.

Jelsema, C.L. and Axelrod, J. (1987). *Proc. Natl. Acad. Sci. USA* **84**, 3623–3627.

Jones, D.R. and Reed, R.R. (1989). *Science* **244**, 790–795.

Jones, T.L.Z., Simonds, W.F., Merendino, J.J., Brann, M.R. and Spiegel, A.M. (1990). *Proc. Natl. Acad. Sci. USA* **87**, 568–572.

Katada, T. and Ui, M. (1979). *J. Biol. Chem.* **254**, 469–479.

Katada, T. and Ui, M. (1982). *J. Biol. Chem.* **257**, 7210–7216.

Kim, D., Lewis, D.L., Graziadei, L., Neer, E.J., Bar-Sagi, D. and Clapham, D.E. (1989). *Nature* **337**, 557–560.

Krupinski, J., Rajaram, R., Lakonishok, M., Benovic, J.L. and Cerione, R.A. (1988). *J. Biol. Chem.* **263**, 12333–12341.

Kurachi, Y., Ito, H., Sugimoto, T., Shimizu, T., Miki, I. and Ui, M. (1989). *Nature* **337**, 555–557.

Landis, C.A., Masters, S.B., Spada, A., Pace, A.M., Bourne, H.R. and Vallar, L. (1989). *Nature* **340**, 692–696.

Lerea, C.L., Somers, D.E. and Hurley, J.B. (1986). *Science* **234**, 77–80.

Levine, M.A., Smallwood, P.M., Moen, P.T. Jr, Helman, L.J. and Ahn, T.G. (1990). *Proc. Natl. Acad. Sci. USA* **87**, 2329–2333.

Longabaugh, J.P., Didsbury, J., Spiegel, A. and Stiles, G.L. (1989). *Mol. Pharmacol.* **36**, 681–688.

Lyons, J., Landis, C.A., Harsh, G., Vallar, L., Grunewald, K., Feichtinger, H., Duh, Q.-Y., Clark, O.H., Kawasaki, E., Bourne, H.R. and McCormick, F. (1990). *Science* **249**, 655–659.

Maguire, M.E., Van Arsdale, P.M. and Gilman, A.G. (1976). *Mol. Pharmacol.* **12**, 335–339.

Maltese, W.A. (1990). *FASEB J.* **4**, 3319–3328.

Matsuoka, M., Itoh, H., Kozasa, T. and Kaziro, Y. (1988). *Proc. Natl. Acad. Sci. USA* **85**, 5384–5388.

Macleod, K.G. and Milligan, G. (1990). *Cell. Signalling* **2**, 139–151.

McKenzie, F.R. and Milligan, G. (1990a). *Biochem. J.* **267**, 391–398.

McKenzie, F.R. and Milligan, G. (1990b). *J. Biol. Chem.* **265**, 17084–17093.

McKenzie, F.R., Adie, E.J. and Milligan, G. (1991). *Biochem. Soc. Trans.* **18**, 81S.

Miller, R.T., Masters, S.B., Sullivan, K.A., Beiderman, B. and Bourne, H.R. (1988). *Nature* **334**, 712–715.

Milligan, G. (1988). *Biochem. J.* **255**, 1–13.

Milligan, G. and Green, A. (1991). *Trends Pharmacol. Sci.* **12**, 207–209.

Milligan, G. and Klee, W.A. (1985). *J. Biol. Chem.* **260**, 2057–2063.

Milligan, G. and Unson, C.G. (1989). *Biochem. J.* **260**, 837–841.

Milligan, G., Unson, C.G. and Wakelam, M.J.O. (1989). *Biochem. J.* **262**, 643–649.

Mochly-Rosen, D., Chang, F.H., Cheever, L., Kim, M., Diamond, I. and Gordon, A.S. (1988). *Nature* **33**, 848–849.

Mumby, S.M., Casey, P.J., Gilman, A.G., Gutowski, S. and Sternweis, P.C. (1990). *Proc. Natl. Acad. Sci. USA* **87**, 5873–5877.

Nagy, L.E., Diamond, I., Casso, D.J., Franklin, C. and Gordon, A.S. (1990). *J. Biol. Chem.* **265**, 1946–1951.

Neer, E.J. and Clapham, D.E. (1988). *Nature* **333**, 129–134.

Northup, J.K., Sternweis, P.C., Smigel, M.D., Schleifer, L.S., Ross, E.M. and Gilman, A.G. (1980). *Proc. Natl. Acad. Sci. USA* **77**, 6516–6520.

Okabe, K., Yatani, A., Evans, T., Ho, Y.-K., Codina, J., Birnbaumer, L. and Brown, A.M. (1990). *J. Biol. Chem.* **265**, 12854–12858.

Rall, T. and Harris, B.A. (1987). *FEBS Lett.* **224**, 365–371.

Reithmann, C., Gierschik, P., Sidiropoulos D., Werdan, K. and Jakobs, K.H. (1989). *Eur. J. Pharmacol.* **172**, 211–221.

Rodbell, M. (1980). *Nature* **284**, 17–22.

Rodbell, M., Krans, H.M.J., Pohl, S. and Birnbaumer, L. (1971). *J. Biol. Chem.* **246**, 1872–1876.

Robishaw, J.D., Kalman, V.K., Moomaw, C.R. and Slaughter, C.A. (1989). *J. Biol. Chem.* **264**, 15758–15761.

Rothenberg, P.L. and Kahn, C.R. (1988). *J. Biol. Chem.* **263**, 15546–15552.

Simonds, W.F., Goldsmith, P.K., Codina, J. Unson, C.G. and Spiegel, A.M. (1989). *Proc. Natl. Acad. Sci. USA* **86**, 7809–7813.

Smrcka, A.V., Helper, J.R., Brown, K.O. and Sternweis, P.C. (1991). *Science*, **251**, 804–807.

Sternweis, P.C. and Robishaw, J.D. (1984). *J. Biol. Chem.* **259**, 13806–13813.

Strassheim, D., Milligan, G. and Houslay, M.D. (1990). *Biochem. J.* **266**, 521–526.

Strathmann, M. and Simon, M.I. (1990). *Proc. Natl. Acad. Sci. USA* **87**, 9113–9117.

Strathmann, M., Wilkie, T.M. and Simon, M.I. (1989). *Proc. Natl. Acad. Sci. USA* **86**, 7407–7409.

Sullivan, K.A., Miller, R.T., Masters, S.B., Beiderman, B., Heideman, W. and Bourne, H.R. (1987). *Nature* **330**, 758–759.

Vallar, I., Spada, A. and Giannattasio, G. (1987). *Nature* **330**, 566–568.

Vallar, L. (1990). *Biochem. Soc. Symp.* **56**, 165–170.

Wakelam, M.J.O., Davies, S.A., Houslay, M.D., McKay, I., Marshall, C.J. and Hall, A. (1986). *Nature* **323**, 173–176.

Watanabe, Y., Imaizumi, T., Misaki, N., Iwakura, K. and Yoshida, H. (1988). *FEBS Lett.* **236**, 372–374.

Yamane, H.K., Farnsworth, C.C., Xie, H., Howald, W., Fung, B.K.-K., Clarke, S., Gelb, M.H. and Glomset, J.A. (1990). *Proc. Natl. Acad. Sci. USA* **87**, 5868–5872.

Zachary, I., Masters, S.B. and Bourne, H.R. (1990). *Biochem. Biophys. Res. Commun.* **168**, 1184–1193.

CHAPTER TWO

Mutations within the Gene Encoding the Stimulatory G-protein of Adenylyl Cyclase as the Basis for Albright Hereditary Osteodystrophy

ALEXANDER MIRIC and MICHAEL A. LEVINE
Johns Hopkins University School of Medicine,
Ross Research Building, Room 1029, 725 North Wolfe
Street, Baltimore, Maryland 21205, USA

1 INTRODUCTION

The term pseudohypoparathyroidism (PHP) describes a heterogeneous syndrome characterized by biochemical hypoparathyroidism (i.e. hypocalcaemia and hyperphosphataemia), elevated plasma levels of parathyroid hormone (PTH), and peripheral unresponsiveness to the biological actions of PTH. Therefore, PHP differs substantially and fundamentally from true hypoparathyroidism, in which deficient synthesis or secretion of biologically active PTH occurs.

The initial description of PHP by Albright and his associates (Albright *et al.*, 1942) focused on the failure of patients with this syndrome to show either a calcaemic or phosphaturic response to parathyroid extract administration. These observations provided the basis for the hypothesis that biochemical hypoparathyroidism in PHP was due not to a deficiency of PTH but rather to resistance of the target organs, bone and kidney, to the biological actions of PTH. PHP was thus the first human disorder (subsequently, many others have been described (Verhoeven and Wilson, 1979)) to be attributed to diminished responsiveness to a hormone by otherwise normal target organs.

G-Proteins
ISBN 0-12-497515-1

2 CLINICAL FEATURES: SIGNS AND SYMPTOMS OF PARATHYROID HORMONE RESISTANCE

Most of the signs and symptoms of PHP are a consequence of reduced concentrations of calcium in blood and extracellular fluid. Ionized calcium, rather than total calcium, is the primary determinant of symptoms in patients with hypocalcaemia. A low extracellular fluid ionized calcium concentration enhances neuromuscular excitability, an effect that is potentiated by hyperkalaemia and hypomagnesemia.

Substantial variation exists among patients both in the absolute level of serum calcium at which symptoms occur and in the severity of symptoms. Subjects with chronic hypocalcaemia sometimes have few, if any, symptoms of neuromuscular irritability despite markedly low serum calcium concentrations. By contrast, acute hypocalcaemia often produces symptoms even when serum calcium concentrations are only mildly depressed. Symptoms of neuromuscular irritability (i.e. tetany) first occur in subjects with PHP at about the eighth year of life; regrettably, these symptoms not infrequently persist for 9–12 years before the correct diagnosis is ascertained (Bronsky et al., 1958; Aurbach and Potts, 1964). Some affected children are relatively (or totally) asymptomatic, and the diagnosis of PHP is established only later in life.

Symptoms of tetany include carpopedal spasm, convulsions, paraesthesias, muscle cramps and stridor. Laryngeal spasm occurs most commonly in young children during episodes of severe hypocalcaemia. In susceptible individuals hypocalcaemia may unmask a previously unsuspected seizure disorder or greatly aggravate known epilepsy. In these patients abnormalities of the electroencephalogram may derive from an underlying seizure disorder or may reflect the nonspecific effects of hypocalcaemia.

Clinical signs of the neuromuscular irritability associated with latent tetany include Chvostek's sign and Trousseau's sign. Chvostek's sign is elicited by tapping the facial nerve anterior to the ear to produce ipsilateral contraction of the facial muscles. Trousseau's sign is present if carpal spasm is induced by pressure ischaemia of nerves of the upper arm during the inflation of a sphygmomanometer above systolic blood pressure for 3–5 min. Slightly positive reactions to either of these tests may occur in normal adults. Importantly, both of these signs can be absent even in patients with severe hypocalcaemia.

Chronic hypocalcaemia is associated with other signs. As with hormonopenic hypoparathyroidism, posterior subcapsular cataracts may develop with long-standing untreated hypocalcaemia and hyper-phosphataemia, and may cause visual impairment. In addition, in approximately 50% of affected subjects radiological examination of the skull will reveal basal ganglion calcification, which may rarely lead to extrapyramidal movement disorders. Dental abnormalities are common in subjects with PHP. The pattern of dental abnormality may help date the onset of hypocalcaemia (Nikiforuk and Frasier, 1979). Ritchie (1965) and Croft *et al.* (1965) have reported an array of such defects, including dentine and enamel hypoplasia, short and blunted roots, and delay (or lack) of tooth eruption. Other features include fragile nails, but moniliasis, which commonly occurs in idiopathic hypoparathyroidism, is not part of PHP. Hypocalcaemia is also associated with prolongation of the corrected QT interval on the electrocardiogram, but other causes of QT prolongation exist. Distinctive olfactory (Weinstock *et al.*, 1986) and gustatory impairments have also been observed in patients with PHP (Henkin, 1968).

3 ALBRIGHT HEREDITARY OSTEODYSTROPHY

In addition to the clinical and biochemical features of hypoparathyroidism, the patients originally described by Albright and his associates (Albright *et al.*, 1942) exhibited a peculiar physical appearance characterized by distinctive skeletal and developmental defects. This unique constellation of somatic characteristics, including round faecies, short stature, obesity, brachydactyly, and heterotopic subcutaneous ossifications, is collectively referred to as Albright hereditary osteodystrophy (AHO; Figure 1). Subsequent to the original description of PHP by Albright and his associates, other patients who exhibited PTH resistance but who lacked features of AHO were identified (Winter and Hughes, 1980). This variant appears to be less common than classic PHP with AHO.

Considerable variability may occur in the clinical expression of AHO, such that all the skeletal and developmental abnormalities characteristic of the syndrome are not necessarily present in each case. Moreover, within AHO families affected members not uncommonly show striking differences in expression of the AHO phenotype. Short stature and obesity are relatively nonspecific physical characteristics, and by themselves are not useful discriminators of the AHO habitus. Brachydactyly, unilateral or bilateral involvement of either hands or feet, is considered an essential criterion of the AHO habitus (Levine and Aurbach, 1989). Brachydactyly in AHO apparently results from premature fusion of the epiphyses of tubular bones. The most commonly shortened bone in AHO is the distal phalanx of the thumb (75%). This is the basis of the thumb sign, often noted in patients with AHO (Ray and Gardner, 1959) or brachydactyly D (stub thumb or murderer's thumb) (Bell, 1951), in which the ratio of the width to the length of the nail is increased. Of the metacarpals, the fourth is most commonly shortened (65%), while the second is the least commonly affected. Asymmetry in shortening is frequent, being found in 42% of cases. Brachydactyly is generally obvious and easily recognized upon physical examination. In other cases, evidence of brachydactyly may be equivocal, and radiological examination will be required for definitive diagnosis.

Brachydactyly can occur in other disorders, e.g. familial brachydactyly, Turner syndrome, and Klinefelter syndrome. The relative

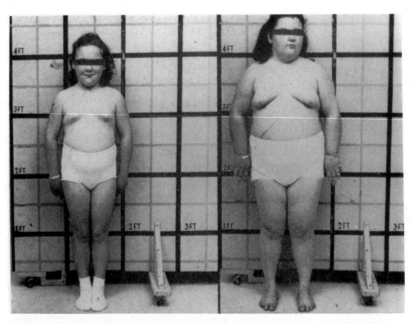

Figure 1. A young girl (left) and her mother (right) with pseudohypoparathyroidism type Ia. Both exhibit features of Albright hereditary osteodystrophy, including short stature, obesity, rounded facies and shortened metacarpals.

shortening and pattern profile of the affected hand bones can be used to distinguish AHO from the various other aetiologies of brachydactyly (Poznanski *et al.*, 1977).

Additional skeletal features of AHO include radius curvus, cubitus valgus, coxa vara, coxa valga, genum varum and genum valgus deformities. The skeletal abnormalities of AHO may not be apparent until a child is four or five years old (Steinbach *et al.*, 1965). At present, the precise basis for the AHO phenotype remains unknown. However, it is clear that expression of AHO occurs independently of the disturbance in calcium metabolism.

4 PATHOGENESIS OF HORMONE RESISTANCE IN PHP

Characterization of the molecular basis for hormone resistance in patients with PHP commenced with the observation that cAMP mediates many of the actions of PTH on kidney and bone, and that administration of biologically active PTH to normal subjects produces a significant increase in the urinary excretion of nephrogenous cAMP (Chase *et al.*, 1969). Patients with PHP type I fail to show an appropriate increase in urinary excretion of both cAMP and phosphate (Chase *et al.*, 1969), whereas subjects with the less common type II form show a normal increase in urinary cAMP excretion but have an impaired phosphaturic response (Drezner *et al.*, 1973). The PTH infusion test remains the most reliable test available for the diagnosis of PHP, and enables distinction between the several variants of the AHO syndrome (Figure 2).

Patients with PHP type I have markedly blunted nephrogenous cAMP responses to exogenous PTH. These observations have suggested that PTH resistance is caused by a defect in the plasma membrane-bound adenylyl cyclase complex that produces cAMP in renal tubule cells. The adenylyl cyclase system is far more complex than originally suspected, consisting of at least three types of proteins embedded in the plasma membrane (review: Gilman, 1987) (Figure 3). Receptors for a large number of hormones and neurotransmitters face the extracellular space and interact with appropriate ligands and drugs. The interaction of such ligands with many types of receptors (R_s) include those for β-adrenergic agonists, PTH, ACTH, gonadotropins, glucagon and many others. Adenylyl cyclase activity is also under inhibitory control by agents such as somatostatin, α_2-adrenergic and muscarinic agonists and opioids. These ligands bind to specific inhibitory receptors (R_i). Receptors communicate with adenylyl cyclase through their interaction with a pair of homologous guanine nucleotide-binding regulatory (G) proteins, one of which (G_s) mediates stimulation of adenylyl cyclase activity, while the other (G_i) is responsible for inhibition (see Chapter 1).

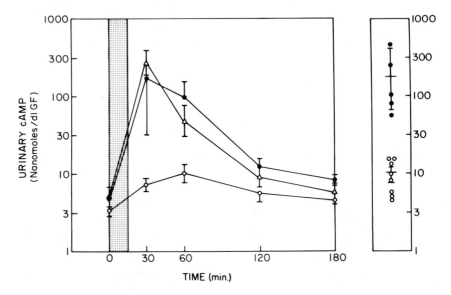

Figure 2. Results of PTH infusion in normal subjects ($\triangle$), AHO patients with PHP type 1 and deficient $G_s\alpha$ activity ($\bigcirc$) and AHO patients with pseudoPHP and deficient $G_s\alpha$ activity ($\bullet$). Left panel: The stippled area represents a 15-min period during which PTH (250U) was infused. Each point represents the mean $\pm$ SEM. GF, glomerular filtration. Right panel: Peak urinary cAMP response to PTH for each AHO patient with PHP type 1 or pseudoPHP. From Levine, M.A., *et al.* (1986).

The G-proteins that regulate activity of adenylyl cyclase are members of a superfamily of signal-transducing GTPase proteins that mediate numerous transmembrane hormonal and sensory transduction processes (reviews: Simon *et al.*, 1991; Neer and Clapham, 1988). The signal-transducing G-proteins share a heterotrimeric structure composed of α, β and γ subunits. The β and γ subunits are tightly associated with each other as a βγ complex, and appear to be functionally interchangeable among at least some of the G-proteins (Cerione *et al.*, 1986; Neer and Clapham, 1988). Nevertheless, the identification of multiple forms of the β (Fong *et al.*, 1986; Gao *et al.*, 1987; Amatruda *et al.*, 1988; Levine *et al.*, 1990) and γ (Fukada *et al.*, 1989; Hurley *et al.*, 1984; Evans *et al.*, 1987; Gautam *et al.*, 1989; Robishaw *et al.*, 1989) polypeptides suggests important structural differences which may define specific functions. The α subunits show the greatest structural diversity, and impart unique functional characteristics to each G-protein. The α subunit contains the guanine nucleotide-binding site, has intrinsic GTPase activity, and is thought to confer functional specificity to each G-protein, allowing discrimination among multiple receptors and effectors.

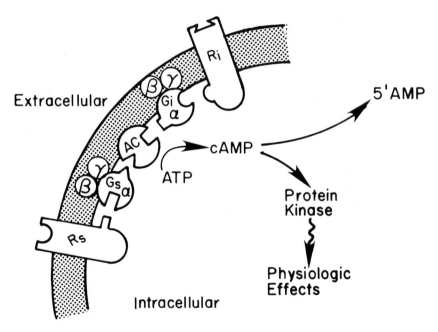

Figure 3. Schematic outline of the adenylyl cyclase system. R_s and R_i, stimulatory and inhibitory receptors; and G_s and G_i, the stimulatory and inhibitory guanine nucleotide-binding regulatory proteins, respectively. AC denotes the catalytic unit of adenylyl cyclase. The subunit structure of the G proteins and their interactions with the catalytic unit are described in the text. From Levine (1990).

Recent studies indicate the existence of at least 16 mammalian genes that encode a superfamily of homologous α subunit proteins (Amatruda *et al.*, 1991). The diversity of α chain structure is further extended by alternative splicing of complex genes, which results in the production of protein variants which may have different receptor and effector specificities. Biochemical studies have elucidated specific roles in signal transduction for nine different mammalian G-protein α chains (Simon *et al.*, 1991). By contrast, molecular cloning, including the use of the polymerase chain reaction, has revealed the existence of cDNAs encoding additional α polypeptides (Strathmann *et al.*, 1989), whose cellular functions at present remain undefined.

Binding of stimulatory hormones, neurotransmitters, or drugs to R_s leads to a conformational change in R_s that enables the receptor to facilitate dissociation of GDP from G_s. Release of GDP permits entry of GTP into the guanine nucleotide-binding site of the α subunit; binding of GTP to G_s subsequently triggers dissociation of the αβ complex from the receptor and separation of the G_sα–GTP complex from the βγ heterodimer. G_sα–GTP stimulates adenylyl cyclase activity and thereby increases synthesis of cAMP. The intracellular accumulation of cAMP produces a biochemical chain reaction that begins with activation of protein kinase A and phosphorylation of specific protein substrates, and ultimately concludes with expression of the physiological response to agonist recognition by the cell (Figure 3). The hydrolysis of GTP to GDP by an intrinsic GTPase promotes reassociation of the G_sα–GDP subunit with the βγ complex, thus restoring G_s to its inactive state. The holoprotein then returns to its orientation with R_s and is ready to participate in another cycle of nucleotide exchange.

Adenylyl cyclase is under 'dual control'. Hormonal inhibition of adenylyl cyclase activity is mediated through activation of G_i by inhibitory receptors (R_i). The interaction of G_i with R_i is similar to that of G_s with R_s, with the exchange of GTP for GDP leading to dissociation of G_i and release of free G_iα–GTP and Gβγ. The mechanism of inhibition of adenylyl cyclase by G_i is not straightforward, however. G_iα–GTP has only a modest inhibitory effect on adenylyl cyclase, but enzyme activity is profoundly reduced by the free Gβγ subunit complex (Katada and Ui, 1982; Katada *et al.*, 1984).

Albright's original description of PHP emphasized PTH resistance as the biochemical hallmark of this disorder (Figure 2). Resistance to PTH alone would be consistent with a defect in the cell surface receptor specific for PTH. However, most patients with PHP type I and AHO display resistance to multiple hormones that activate adenylyl cyclase (Levine *et al.*, 1983), and express additional abnormalities, such as mental retardation (Farfel and Friedman, 1986), and impaired olfaction (Weinstock *et al.*, 1986). The presence of these multiple defects is consistent with a generalized abnormality that impairs production of cAMP in all tissues. These AHO patients, referred to as PHP type Ia, have an approximately 50% reduction in $G_s\alpha$ activity in plasma membranes from multiple cell types (Levine and Aurbach, 1989; Figure 4). In contrast, membrane levels of $G_i\alpha$ are normal (Downs *et al.*, 1985). A generalized deficiency of $G_s\alpha$ may provide one mechanism to explain the reduced ability of hormones and neurotransmitters to activate adenylyl cyclase in multiple tissues, and thereby lead to widespread hormone resistance.

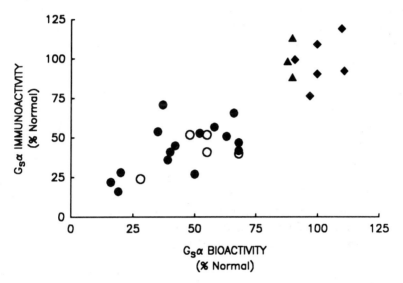

Figure 4. Comparative analysis of cell membrane $G_s\alpha$ immunoactivity and bioactivity. Cell membranes from 20 patients with $G_s\alpha$ deficiency (●, PHP type Ia; ○, pseudoPHP); three patients with PHP type Ib (▲), and six normal subjects (◆) were analysed. The relative levels of immunoactive $G_s\alpha$ protein were determined by quantitative immunoblot analysis and expressed as a percentage of values obtained from a group of normal subjects. $G_s\alpha$ bioactivity (on the abscissa) was determined by a complementation assay and expressed as a percentage of the mean activity of control membranes. From Patten, J.L. and Levine, M.A. (1990).

Surprisingly, some subjects with deficient $G_s\alpha$ activity and AHO do not manifest biochemical evidence of hormone resistance (Figure 2). Albright termed this normocalcaemic variant of AHO 'pseudopseudo-hypoparathyroidism' (pseudoPHP) to call attention to the physical similarity (AHO) yet metabolic dissimilarity (normal PTH responsiveness) of this disorder to PHP (Albright *et al.*, 1952). PseudoPHP is genetically related to PHP type I. Early clinical observations of AHO kindreds in which several affected members had only AHO (i.e. pseudoPHP) while others had PTH resistance as well (i.e. PHP type I) first suggested that the two disorders might reflect variability in expression of a single genetic lesion (Fitch, 1982; Kinard *et al.*, 1979). Further support for this view derives from recent studies indicating that, within a given kindred, subjects with either pseudoPHP or PHP type Ia have equivalent functional $G_s\alpha$ deficiency (Levine *et al.*, 1986, 1988) and identical mutations in the gene encoding $G_s\alpha$ (see below). We therefore suggest that the term AHO be used to simplify description of this syndrome, and to acknowledge the common clinical and biochemical characteristics that patients with PHP type Ia and pseudoPHP share.

5 MOLECULAR PATHOPHYSIOLOGY OF $G_s\alpha$ DEFICIENCY IN ALBRIGHT HEREDITARY OSTEODYSTROPHY

$G_s\alpha$ deficiency in AHO has emerged as an important human paradigm for defining the nature and functional consequences of impaired signal transduction through the adenylyl cyclase effector system. Several aspects of this inherited disorder make it particularly amenable to detailed molecular analyses. Our knowledge of normal $G_s\alpha$ protein (Gilman, 1987), mRNA (Robishaw *et al.*, 1986; Bray *et al.*, 1986), and gene (Kozasa *et al.*, 1988) provides a basis for the precise delineation of the abnormalities present in $G_s\alpha$-deficient patients. The isolation and analysis of mutant $G_s\alpha$ genes and their products is simplified because the gene is expressed in a wide variety of tissues.

The apparent autosomal dominant transmission of $G_s\alpha$ deficiency in patients with AHO (Levine *et al.*, 1986) has led to the speculation that the primary defect in this disorder involves the $G_s\alpha$ structural gene. $G_s\alpha$ is encoded by the GNAS1 gene, a single, 13-exon gene spanning 20 kb (Kozasa *et al.*, 1988). An alternative splicing model has been proposed to explain the generation of four different $G_s\alpha$ mRNAs, each approximately 1.85 kb in size, which give rise to the 45-kDa and 52-kDa forms of $G_s\alpha$ protein. The inclusion of exon 3 results in the insertion of 15 additional amino acid residues; further diversity derives from the use of alternative splice acceptor sites in exon 4. This process results in the generation of two long (52-kDa) and two short (45-kDa) $G_s\alpha$ molecules that differ only by the addition of a single serine residue (Kozasa *et al.*, 1988; Robishaw *et al.*, 1986).

Patients with $G_s\alpha$ deficiency have a variety of genetic mutations. Indeed, AHO patients may have reduced (Carter *et al.*, 1987; Levine *et al.*, 1988) or normal (Levine *et al.*, 1988) levels of $G_s\alpha$ mRNA, consistent with the presence of genetic heterogeneity in the molecular pathophysiology of this disorder. Additional support for this notion derives from recent immunochemical analyses of $G_s\alpha$ protein in erythrocyte and fibroblast membranes from AHO subjects with $G_s\alpha$ deficiency. Erythrocyte membranes from affected members of one AHO kindred with normal levels of $G_s\alpha$ mRNA contained a putative abnormal form of $G_s\alpha$ that lacked reactivity with antisera that recognize epitopes encoded by the first two exons of the $G_s\alpha$ gene (Patten *et al.*, 1990). Direct analysis of the mutant $G_s\alpha$ gene coding for the abnormal form of $G_s\alpha$ in the two affected patients of this AHO family revealed an A→G transition at position $+1$ within the ATG initiator methionine codon, blocking initiation of translation at the normal site. The next downstream ATG codon is inframe and corresponds to the methionine residue at amino acid position 60; the observation that this abnormal $G_s\alpha$ protein is recognized by a C-terminal-directed anti-$G_s\alpha$ peptide antiserum is consistent with the notion that the abnormal mRNA produces a defective $G_s\alpha$ protein that lacks the first 59 amino acids of the N-terminus. Recent evidence suggests that the N-terminus plays a critical role in the functional capacity of the $G_s\alpha$ protein. Binding of the $\beta\gamma$ dimer to the α subunit requires the N-terminus (Neer *et al.*, 1988). Without this interaction, receptor-catalysed guanine nucleotide exchange would be much less efficient. Close proximity of the N- and C-termini of $G_s\alpha$ appears necessary for the key regulatory domains in ligand–receptor and $\beta\gamma$ interactions within the tertiary structure of the $G_s\alpha$ molecule to be accessible. This multi-component regulation complex facilitating cellular signal transduction would not function at maximal capacity without the participation of the α subunit's N-terminus. Thus, it is likely that the abnormal $G_s\alpha$ molecule in these affected patients leads to reduced $G_s\alpha$ bioactivity and AHO.

Immunochemical studies of cell membranes from affected members of eight additional AHO kindreds have failed to disclose other abnormal forms of $G_s\alpha$ protein (Patten *et al.*, 1990). Fibroblast membranes from patients with AHO who had reduced or normal levels of $G_s\alpha$ mRNA contained both the 45- and 52-kDa forms of the $G_s\alpha$ protein in quantities that were significantly less ($52 \pm 6\%$ (mean $\pm$ SEM, $n = 8$) for AHO patients with reduced quantities of $G_s\alpha$ mRNA, and $35 \pm 19\%$ (mean $\pm$ range, $n = 2$) for AHO patients with normal quantities of $G_s\alpha$ mRNA, percentage of control values) than those present in membranes from normal subjects. Similar reductions were found in the level of the

45-kDa form of $G_s\alpha$ in erythrocyte membranes from all AHO patients studied ($40 \pm 4\%$, mean $\pm$ SEM, of control values). There was a significant ($p < 0.05$) correlation between levels of $G_s\alpha$ immunoactivity and bioactivity in these patients (Patten and Levine, 1990; Figure 4).

Restriction endonuclease analysis of genomic DNA from subjects with AHO has failed to reveal large deletions or rearrangements of the $G_s\alpha$ gene that could account for altered gene expression. Detection and localization of small mutations within the GNAS1 gene in subjects with AHO is impractical by complete sequence determination and improbable on the basis of restriction endonuclease vulnerability. Thus, efforts to elucidate the molecular basis for AHO in additional families have focused on application of three techniques that help to overcome these problems: DNA amplification by the polymerase chain reaction (PCR), denaturing gradient gel electrophoresis (DGGE), and direct sequencing of the PCR product. A recently developed technique, DGGE satisfies many of the requirements for a screening method that can rapidly identify point mutations in defined regions of DNA. DGGE is capable of detecting single base substitutions in DNA fragments of 100–1000 bp in length on the basis of sequence-dependent melting properties of double-stranded nucleic acid molecules (Fischer and Lerman, 1983; Myers et al., 1985; Myers and Maniatis, 1986). Double-stranded DNA undergoes an abrupt transition from the totally helical state to a partially melted state when it migrates to a specific denaturant concentration in the gel; the denaturant concentration required for this transition is dependent upon the nucleotide sequence. DNA fragments differing by as little as a single base substitution will melt at slightly different denaturant concentrations because of differences in the stacking interactions between adjacent bases in each DNA strand. Accordingly, wild-type and mutant alleles of amplified DNA fragments will migrate differently within the electrophoretic gel. In practice genomic DNA from subjects with $G_s\alpha$ deficiency is amplified by PCR using intron oligonucleotide primers flanking the 13 exons of the $G_s\alpha$ gene. To increase the sensitivity of DGGE, one oligonucleotide from each primer pair is synthesized with a 40-nucleotide GC-rich sequence at the 5' end. Addition of a 40-bp GC-rich sequence, designated a GC-clamp, to one end of amplified DNA fragments can enable DGGE to detect nearly 100% of single base changes (Sheffield et al., 1989). By comparison to the GC-rich clamp, the test segment is relatively GC-poor, and is the first melting domain. The DNA fragments are screened for mutations by observing differences in their migration on formamide/urea gradient denaturing polyacrylamide gels.

Because AHO is an autosomal dominant disorder, an affected subject will have one normal (wild type) and one abnormal $G_s\alpha$ allele.
Thus, each amplified DNA sample in which a defective $G_s\alpha$ allele is
present will demonstrate *two homoduplex bands* and two additional
bands representing *heteroduplexes* formed during the PCR amplification (Figure 5). Abnormal DNA fragments from AHO patients can be
isolated individually from polyacrylamide gel slices and sequenced.

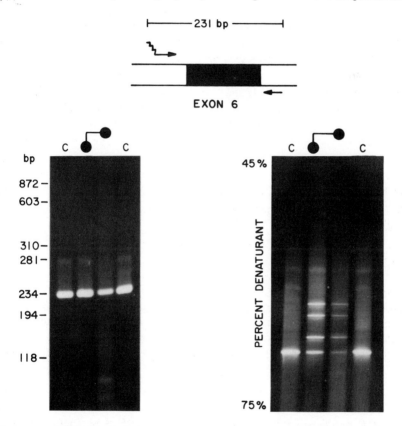

Figure 5. Analysis of $G_s\alpha$ gene amplification products. Left panel: Exon 6
products (231 bp) were analysed by standard 5% polyacrylamide gel electrophoresis. Lanes labelled C indicate control subjects and lanes beneath partial
pedigree indicate subjects with AHO and $G_s\alpha$ deficiency shown in Figure 2.
Location of the molecular weight markers is indicated on the left. Right panel:
Denaturant gradient gel electrophoresis of the same amplification products.
Single bands in lanes marked C represent homoduplex DNA containing the
normal gene sequence. DNA in lanes beneath the partial pedigree shows
presence of two homoduplexes and two more slowly migrating heteroduplexes—a pattern consistent with the presence of a heterozygous mutation.

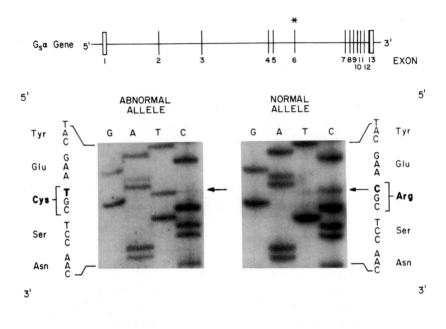

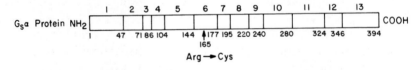

Arg → Cys

Figure 6. Direct sequence analysis of exon 6 fragment of the $G_s\alpha$ gene. Homoduplex bands containing the normal and mutant amplification products were cut from the denaturing gel, re-amplified, and directly sequenced. Sequence of the more rapidly migrating DGGE homoduplex (Figure 5) contained the normal $G_s\alpha$ gene sequence. By contrast, sequence of the more slowly migrating DGGE homoduplex (Figure 6) revealed a C→T transition. This mutation produces an arginine to cysteine substitution at amino acid 165.

DGGE analysis of $G_s\alpha$ gene fragments amplified by PCR has facilitated identification of three additional types of mutations. Levine *et al.* (unpublished data) have identified missense mutations in subjects with AHO and have proposed this mechanism as one basis for deficient $G_s\alpha$ activity in AHO. Affected members of one AHO kindred, shown in Figure 2, contain a single C→T transition in exon 6 that results in a missense mutation changing the codon for Arg to Cys at amino acid position 165 of the predicted abnormal $G_s\alpha$ protein (Figure 6). In a second family a nucleotide substitution CTG→CCG results in replacement of Leu by Pro at codon 99 of the $G_s\alpha$ molecule. In both cases amphipathic and secondary structure analysis by the Chou Fasman algorithm of the primary structure of the abnormal $G_s\alpha$ proteins predicts disruption of α-helical conformation in domains that are predicted to be involved in binding of the $G_s\alpha$ chain to adenylyl cyclase.

Weinstein *et al.* (1990) have described a G→C substitution of the first base of intron 10 at the donor splice junction bordering the 3′ end of exon 10. In a second family described by Levine *et al.* (unpublished data), an A→G transition was identified 12 bases from the 3′ terminus of intron 3. This mutation generates a consensus sequence for the 3′ splice site for an intron, CAG, and predicts an 11-bp addition to the mRNA that causes a frameshift and a premature stop codon. These splice junction mutations are predicted to result in abnormal RNA processing, and could explain the reduced steady-state $G_s\alpha$ mRNA levels found in affected members of these kindreds.

Finally, nucleotide deletions that produce coding frameshift mutations have also been identified in subjects with AHO. Weinstein *et al.* (1990) described a second AHO kindred in which affected subjects exhibited a single base deletion in exon 10 at codon 272. A four-base deletion in exon 4 has recently been identified by Miric and Levine (unpublished data) in affected subjects from another AHO family. In both cases, the deletions cause frameshift mutations which result in premature termination of translation of the $G_s\alpha$ mRNA.

In all cases reported to date, nucleotide sequences in the corresponding site of the other $G_s\alpha$ gene allele have been normal. The findings in all reported families are consistent with the autosomal dominant inheritance of AHO by a heterozygous mutation in the $G_s\alpha$ gene and reaffirm that a variety of inherited molecular defects result in the AHO phenotype and functional deficiency of $G_s\alpha$. Moreover, in each family studied, identical gene defects have been found in subjects with either PHP type Ia or pseudoPHP that explain the equivalent reductions in functional $G_s\alpha$ activity in these patients. These observations further emphasize the genetic and biochemical association between these two variants of the AHO syndrome, but leave unanswered the intriguing question of why $G_s\alpha$ deficiency does not cause hormone resistance in all affected subjects. One speculation is that $G_s\alpha$ deficiency is necessary for expression of AHO, but is not sufficient for expression of the complete disorder (i.e. hormone resistance). In this model the effects of $G_s\alpha$ deficiency could be modified by additional, non-allelic genes, that encode other proteins important for cAMP metabolism. Patients with hormone resistance might express gene products that aggravate $G_s\alpha$ deficiency (e.g. a cAMP phosphodiesterase with high activity) and thereby result in reduced accumulation of cAMP after hormone stimulation. Alternatively, patients with $G_s\alpha$ deficiency who lack hormone resistance might express gene products that mitigate $G_s\alpha$ deficiency (e.g. a cAMP phosphodiesterase with low activity) and thereby result in normal accumulation of cAMP after hormone stimulation.

Cloning and sequencing have been effective tools to establish the molecular basis for functional deficiency of $G_s\alpha$ in patients with AHO. However, these approaches have given us limited insight into the basis for hormone resistance. Application of homologous recombination should provide ways to inactivate single $G_s\alpha$ genes and allow assessment of the functional consequences of $G_s\alpha$ deficiency in transgenic mice. Similar studies of other genes in transgenic mice have proven to be highly informative, and may be extended to the model of AHO to elucidate the complex interactions of $G_s\alpha$ within the hormone-responsive adenylyl cyclase signal transduction cascade.

REFERENCES

Albright, F., Burnett, C.H., Smith, P.H. and Parson, W. (1942). *Endocrinology* **30**, 922–932.

Albright, F., Forbes, A.P. and Henneman, P.H. (1952). *Trans. Assoc. Am. Physicians* **65**, 337–350.

Amatruda, T.T., Gautam, N., Fong, H.K., Northup, J.K. and Simon, M.I. (1988). *J. Biol. Chem.* **263**, 5008–5011.

Amatruda, T.T., Steele, D.A., Slepak, V.Z. and Simon, M.I. (1991). *Proc. Natl. Acad. Sci. USA* **88**, 5587–5591.

Aurbach, G.D. and Potts, J.T. Jr (1964). In: *Advances in Metabolic Disorders*, Vol. 1 (Levine, R. and Luft, R., eds), p. 45. Academic Press, New York.

Bell, J. (1951). In: *The Treasury of Human Inheritance*, Vol. 5 (Penrose, L.S., ed.), p. 1. University Press, Cambridge.

Bray, P., Carter, A., Simons, C., Guo, V., Puckett, C., Kamholz, J., Spiegel, A. and Nirenberg, N. (1986). *Proc. Natl. Acad. Sci. USA* **83**, 8893–8897.

Bronsky, D., Kushner, D.S., Dubin, A. and Snapper, I. (1958). *Medicine* **37**, 317–352.

Carter, A., Bardin, C., Collins, R., Simons, C., Bray, P. and Spiegel, A. (1987). *Proc. Natl. Acad. Sci. USA* **84**, 7266–7269.

Cerione, R.A., Staniszewski, C., Gierschik, P., Codina, J., Somers, R.L., Birnbaumer, L., Spiegel, A.M., Caron, M.G. and Lefkowitz, R.J. (1986). *J. Biol. Chem.* **261**, 9514–9520.

Chase, L.R., Melson, G.L. and Aurbach, G.D. (1969). *J. Clin. Invest.* **48**, 1832–1844.

Croft, L.K., Witkop, C.J. Jr and Glas, J.E. (1965). *Oral Surg.* **20**, 758–770.

Downs, R.W., Sekura, R.D., Levine, M.A. and Spiegel, A.M. (1985). *J. Clin. Endocrinol. Metab.* **61**, 351–354.

Drezner, M.K., Neelon, F.A. and Lebovitz, H.E. (1973). *N. Engl. J. Med.* **289**, 1056–1060.

Evans, T., Fawzi, A., Fraser, E.D., Brown, M.L. and Northup, J.K. (1987). *J. Biol. Chem.* **262**, 176–181.

Farfel, Z. and Friedman, E. (1986). *Ann. Intern. Med.* **105**(2), 197–199.

Fischer, S.G. and Lerman, L.S. (1983). *Proc. Natl. Acad. Sci. USA* **80**, 1579–1583.

Fitch, N. (1982). *Am. J. Med. Genet.* **11**, 11–29.

Fong, H.K.W., Hurley, J.B., Hopkins, R.S., Miake-Lye, R., Johnson, M.S., Doolittle, R.F. and Simon, M.I. (1986). *Proc. Natl. Acad. Sci. USA* **83**, 2162–2166.

Fukada, Y., Ohguro, H., Saito, T., Yoshizawa, T. and Akino, T. (1989). *J. Biol. Chem.* **264**, 5937–5943.

Gao, B., Gilman, A.G. and Robishaw, J.D. (1987). *Proc. Natl. Acad. Sci. USA* **84**, 6122–6125.

Gautam, N., Baetscher, M., Aebersold, R. and Simon, M.I. (1989). *Science* **244**, 971–974.

Gilman, A.G. (1987). *Annu. Rev. Biochem.* **56**, 615–649.

Henkin, R.I. (1968). *J. Clin. Endocrinol. Metab.* **28**, 624–628.

Hurley, J.B., Fong, H.K.W., Teplow, D.B., Dreyer, W.J. and Simon, M.I. (1984). *Proc. Natl. Acad. Sci. USA* **81**, 6948–6952.

Katada, T. and Ui, M. (1982). *Proc. Natl. Acad. Sci. USA* **79**, 3129–3133.

Katada, T., Northrup, J.K., Bokoch, G.M., Ui, M. and Gilman, A.C. (1984). *J. Biol. Chem.* **259**, 3578–3585.

Kinard, R.E., Walton, J.E. and Buckwalter, J.A. (1979). *Arch. Intern. Med.* **139**, 204–207.

Kozasa, T., Itoh, H., Tsukamoto, T. and Kaziro, Y. (1988). *Proc. Natl. Acad. Sci. USA* **85**, 2081–2085.

Levine, M.A. (1990). In: *Primer on the Metabolic Bone Diseases and Disorders of Mineral Metabolism* (Favus, M.J., ed.), 1st edn, p. 131. William Byrd Press, Richmond, Virginia.

Levine, M.A. and Aurbach, G.D. (1989). In: *Endocrinology* Vol. 2 (DeGroot, L.J., ed.), pp. 1065–1079. W. B. Saunders Company, Philadelphia.

Levine, M.A., Downs, R.W. Jr, Moses, A.M., Breslau, N.A., Marx, S.J., Marx, S.R., Lasker, R.D., Rizzoli, R.E., Aurbach, G.D. and Spiegel, A.M. (1983). *Am. J. Med.* **74**, 545–556.

Levine, M.A., Jap, T.-S., Mauseth, R.W., Downs, R.W. and Spiegel, A.M. (1986). *J. Clin. Endocrinol. Metab.* **62**, 497–502.

Levine, M.A., Ahn, T.G., Klupt, S.F., Kaufman, K.D., Smallwood, P.M., Bourne, H.R., Sullivan, K.A. and Van Dop, C. (1988). *Proc. Natl. Acad. Sci. USA* **85**, 617–621.

Levine, M.A., Smallwood, P.M., Moen, P.T. Jr, Helman, L.J. and Ahn, T.G. (1990). *Proc. Natl. Acad. Sci. USA* **87**, 2329–2333.

Myers, R.M. and Maniatis, T. (1986). In: *Molecular Biology of Homo Sapiens— Cold Spring Harbor Symposia on Quantitative Biology*, v. 51, Cold Spring Harbor Laboratory, Cold Spring Harbor, NY, pp. 275–284.

Myers, R.M., Fischer, S.G., Maniatis, T. and Lerman, L.S. (1985). *Nucleic Acids Res.* **13**, 3111–3129.

Neer, E.J. and Clapham, D.E. (1988). *Nature* **333**, 129–134.

Neer, E.J., Pulsifer, L. and Wolf, L.G. (1988). *J. Biol. Chem.* **263**, 8996–9000.

Nikiforuk, G. and Frasier, D. (1979). *Metab. Bone Dis.* **2**, 17.

Patten, J.L. and Levine, M.A. (1990). *J. Clin. Endocrinol. Metab.* **71**(5), 1208–1214.

Patten, J.L., Johns, D.R., Valle, D., Eil, C., Gruppuso, P.A., Steele, G., Smallwood, P.M. and Levine, M.A. (1990). *N. Engl. J. Med.* **322**, 1412–1419.

Poznanski, A.K., Werder, E.A., Giedion, A., Martin, A. and Snow, H. (1977). *Radiology*, **123**, 707–718.

Ray, E.W. and Gardner, L.I. (1959). *Am. J. Dis. Child* **96**, 599–601.

Ritchie, G.M. (1965). *Arch. Dis. Child.* **40**, 565–572.

Robishaw, J.D., Smigel, M.D. and Gilman, A.G. (1986). *J. Biol. Chem.* **261**, 9587–9590.

Robishaw, J.D., Kalman, V.K., Moomaw, C.R. and Slaughter, C.A. (1989). *J. Biol. Chem.* **264**, 15758–15761.

Sheffield, V.C., Cox, D.R., Lerman, L.S. and Myers, R.M. (1989). *Proc. Natl. Acad. Sci. USA* **86**, 232–236.

Simon, M.I., Strathmann, M.P. and Gautam, N. (1991). *Science* **252**, 802–808.

Steinbach, H.L., Rudhe, U., Jonsson, M. and Young, D.A. (1965). *Radiology* **85**, 670–676.

Strathmann, M., Wilkie, T.M. and Simon, M.I. (1989). *Proc. Natl. Acad. Sci. USA* **86**, 7407–7409.

Verhoeven, G.F.M. and Wilson, J.D. (1979). *Metabolism* **28**, 253–289.

Weinstein, L.S., Gejman, P.V., Friedman, E., Kadowaki, T., Collins, R.M., Gershon, E.S. and Spiegel, A.M. (1990). *Proc. Natl. Acad. Sci. USA* **87**, 8287–8290.

Weinstock, R.S., Wright, H.N., Spiegel, A.M., Levine, M.A. and Moses, A.M. (1986). *Nature* **322**, 635–636.

Winter, J.S.D. and Hughes, I.A. (1980). *Can. Med. Assoc. J.* **123**, 26–31.

CHAPTER THREE

GTP-Binding Proteins and Cardiovascular Disease

KAZUSHI URASAWA and PAUL A. INSEL
Department of Pharmacology, University of California at San Diego, La Jolla, CA 92093, USA

1 INTRODUCTION

The role of GTP-binding protein (G-protein)-linked pathways in mediating response to a wide variety of neurohormonal signals is now well established (Birnbaumer *et al.*, 1990). Less well defined is the possible role of changes in these pathways in disease processes, such as those that involve the cardiovascular system, the topic of this chapter.

In the cardiovascular system, which for the purpose of this chapter involves the heart and blood vessels, a large number of G-protein-linked receptors have been identified (Table 1). The vast majority of the experimental studies on changes in disease have emphasized adrenergic receptors (AR), which recognize and respond to circulating and sympathetic neuronally released catecholamines, and muscarinic cholinergic receptors, which respond to parasympathetic neuronally released acetylcholine. In turn, a rapidly growing body of data has focused on the G-proteins which couple those receptors to effector molecules (Freissmuth *et al.*, 1989). In this chapter, we will emphasize post-receptor changes that occur in cardiovascular disease.

It is important to realize that assays for G-proteins and G-protein-linked effector molecules are not yet (i.e. in early 1991) standardized and optimized between laboratories. This problem contributes to the relative dearth of information as well as to some of the inconsistent results that have been published. Until quite recently, the typical approach for examining possible changes in G-proteins was to conduct indirect assays of the proteins, such as changes in agonist binding to

G-Proteins
ISBN 0-12-497515-1

Table 1. G-protein-coupled receptors in the cardiovascular system

Receptor	G-protein	Effector linkage	Reference
$\beta_1(\beta_2)$-adrenergic	G_s	AC activation	Sutherland et al. (1962)
		Ca^{2+} channel activation	Yatani et al. (1987b)
		Na^+ channel inhibition	Schubert et al. (1989)
α_1-adrenergic	$G_?$	PLC activation	Brown et al. (1985)
α_2-adrenergic	G_i (or G_o)	?	Nichols et al. (1988)
M_2-muscarinic	G_i	AC inhibition	Mattera et al. (1985)
	G_k, G_i	K^+ channel activation	Yatani et al. (1987a)
D_1-dopamine	G_s	AC activation	Goldberg (1972)
A_1-adenosine	G_i	AC inhibition	Hazeki et al. (1983)
	G_k, G_i	K^+ channel activation	Kurachi et al. (1986)
H_2-histamine	G_s	AC activation	Klein and Levey (1971)
Angiotensin II	$G_?$	PLC activation	Alexander et al. (1985)
Serotonin	$G_?$	PLC activation	Bruns and Marmé (1987)
Glucagon	G_s	AC activation	Levey and Epstein (1969) Murad and Vaughan (1969)
VIP	G_s	AC activation	Chatelain et al. (1980)
Somatostatin	G_i	AC inhibition	Vinicor et al. (1977)
ANF	G_i	AC inhibition	Anand-Srivastava et al. (1987)
Bradykinin	$G_?$	PLC activation	Lambert et al. (1986)
	$G_?$	PLA_2 activation	Hong and Deykin (1982)
Neuropeptide Y	G_i	AC inhibition	Kassis et al. (1987)
	G_k, G_i	K^+ channel activation	Birnbaumer et al. (1990)

VIP, vasoactive intestinal peptide; ANF, atrial natriuretic factor; AC, adenylyl cyclase; PLC, phospholipase C; PLA_2, phospholipase A_2.

receptors (usually determined in the absence and presence of guanine nucleotides) or in adenylyl cyclase activity in response to a variety of stimulants (hormonal agonists, non-hydrolysable guanine nucleotides, AlF_4^-, forskolin, cholera toxin, pertussis toxin, Mn^{2+}) and to compare the patterns of response observed in normal or diseased tissue. We are just now in the era where these types of studies are being supplanted by more direct studies that involve quantitation of G-proteins by antibodies, toxin-catalysed ADP-ribosylation, or, in some cases, functional reconstitution into G-protein-deficient membranes. Our bias is that use of a specific antibody combined with techniques such as quantitative immunoblotting or ELISA will provide the most precise information regarding the *amount* of a particular G-protein but that ideally one would also conduct studies, such as reconstitution, to provide an assessment of *functional state* of the protein. (As discussed elsewhere, toxin-catalysed ADP-ribosylation, especially in the case of cholera toxin, may not be as precise as use of antibodies for quantitation of the amount of G-proteins (Ransnas and Insel, 1988, 1989).) As will be discussed, examples have been described in which diseased tissue can have an unaltered amount of G-protein but the protein does not appear to be fully functional, perhaps due to covalent modification or other factors.

Assessment of changes in effector molecules is at an even more primitive stage than is assessment of G-proteins. Direct methods (e.g. antibodies) for quantitation of molecular entities such as the catalyst of adenylyl cyclase or the G-protein-linked phospholipases are not yet available. Indirect approaches, such as assessment of enzymic activity or binding of activators, e.g. labelled forskolin (which binds to the catalyst of adenylyl cyclase), may provide indirect information regarding effector molecules. In this regard, it has recently been proposed that agonist-promoted or cholera toxin-promoted binding of forskolin to the catalyst of adenylyl cyclase can provide an estimate of the number of G-protein–catalyst complexes formed in target cells (Barber, 1988; Alousi *et al.*, 1991). This approach has not yet been validated in relevant cardiovascular tissues.

Although we do not yet know all of the G-proteins that are found in cardiovascular tissues, several have been definitively identified. These include two forms of G_s, 52-kDa and 45-kDa forms (Murakami and Yasuda, 1986); at least two forms of G_i, G_i2 and G_i3; and G_o (Luetje *et al.*, 1988; Urasawa *et al.*, 1990a). It is now known that G_s in the heart can link to at least two distinct effector molecules; the catalyst of adenylyl cyclase and voltage-sensitive L-type Ca^{2+} channel (Yatani *et al.*, 1987b). Since cAMP generated by adenylyl cyclase acts via cAMP-dependent protein kinase (PKA) to regulate the phosphorylation state, and in turn the function of Ca^{2+} channels in the heart, nature has evolved both an indirect and direct way for agonists acting via G_s to regulate Ca^{2+} channel activity. In addition, limited data suggest that G_s may also link to Na^+ channels in the heart (Schubert et al., 1989). Most previous work related to cardiovascular disease has emphasized the linkage of G_s to adenylyl cyclase and may have ignored other physiologically (and pathophysiologically) important aspects of this G-protein pathway. Like G_s, G_i2 in the heart is also able to couple to at least two specific effector pathways: the inhibition of adenylyl cyclase activity and an inward rectifying K^+ channel (Yatani et al., 1987a). The role that each of these may play in disease settings has not been well defined, even though a number of pathological settings have altered expression of G_i. G_o is present in heart but its precise functional role is, as yet, unknown. Other functional activities that presumably are linked to G-proteins remain to be determined: activation of phospholipase C (presumably via what has been termed G_p but which may be more accurately termed G_q) (Simon et al., 1991; Smrcka et al., 1991), activation of phospholipase A_2, and perhaps regulation of other types of ion channels and transport pathways, such as the Na^+/H^+ exchanger.

In this chapter we will focus our presentation on four types of cardiovascular disorders: cardiomyopathy/congestive heart failure, myocardial ischaemia, hypertension, and complications of drug therapy. This selection was made mainly due to space limitations and the relatively small data base in other areas. The difficulty in obtaining heart samples from patients has been a major obstacle for biochemical and molecular biological approaches in human cardiovascular disease. For this reason, much more information is available from studies conducted with animal models than with human subjects, except for those studies of congestive heart failure in explanted hearts from recipients of heart transplantation. Nevertheless, since the principal goal of these types of studies is to gain a better understanding of the underlying pathophysiology in man, we have attempted to focus our presentation toward human disease.

2 SIGNALLING ABNORMALITIES IN CARDIOVASCULAR DISEASES

2.1 Cardiomyopathy

Idiopathic cardiomyopathy in humans is classified into three categories: dilated cardiomyopathy, hypertrophic cardiomyopathy, and restrictive cardiomyopathy. Of these, idiopathic dilated cardiomyopathy

(DCM) has the worst prognosis and has been the most extensively studied.

2.1.1 Animal models

Since the first report by Bajusz *et al.* (1966), cardiomyopathic Syrian hamsters have been widely used as model animals to study the bio-chemical abnormalities in cardiomyopathy. Because of characteristic histological changes, calcium overload has been implicated as a major mechanism of muscle degeneration in this model (Jasmin and Pros-chek, 1984). For this reason, much experimental work has emphasized cardiac handling of Ca^{2+}, including assessment of sarcolemmal Ca^{2+} channels and Ca-ATPase in sarcoplasmic reticulum, and evaluation of the effectiveness of Ca^{2+} channel blocker therapy (Gertz, 1972; Jasmin and Solymoss, 1975; Finkel *et al.*, 1986; Wagner *et al.*, 1986; Whitmer *et al.*, 1988). Two different strains of cardiomyopathic Syrian hamsters have been commonly used: BIO14.6 and BIO53.58.

Development of cardiac lesions in BIO14.6 can be divided into four phases (Gertz, 1972): (1) prenecrotic (before 30 days), (2) focal necrotic (30–90 days), (3) hypertrophic (after 90 days), and (4) terminal. Studies of plasma membrane receptors and G-proteins have been undertaken at these different phases of disease. For example, Kagiya *et al.* (1985) reported that β-adrenergic receptor (βAR) density was increased at the focal necrotic stage, then declined during the hypertrophic stage, while an increase of αAR persisted during the entire course of disease. These workers speculated that an increased αAR stimulation might play a role in pathogenesis. Investigation at a much earlier stage of disease was conducted by Kobayashi *et al.* (1987). Although these workers found a significant increase of βAR and αAR (α adrenergic receptor) at the focal necrotic stage, these alterations were not observed during the prenecrotic stage. On the other hand, Ca^{2+} channels, detected by labelled dihydropyridine binding, were significantly increased even at the prenecrotic stage. Furthermore, early administration of the Ca^{2+} channel antagonist verapamil prevented the fibrosis and calcification of the myocardium of BIO14.6. These results led the authors to conclude that an increased expression of voltage-sensitive Ca^{2+} chan-nels during the prenecrotic stage may participate in intracellular Ca^{2+} accumulation, followed by free radical overproduction, and sub-sequent cell damage.

Some information has been reported regarding alterations in G-proteins in BIO14.6 hamsters. A functional defect of G_s in cardiac and skeletal muscle of BIO14.6 was reported by Kessler *et al.* (1989). Assay of G_s by reconstitution using S49 cyc$^-$ cell membranes revealed reduced G_s activity in BIO14.6 at the prenecrotic stage, even though there was no difference (compared to controls) in the quantity of G_s, as assessed by cholera toxin-catalysed ADP-ribosylation or by immunoblot analysis. Additional data indicated a decrease in $G_s\alpha$ mRNA expression between BIO14.6 and control hamsters. Subsequent studies indicated that the level of $G_s\alpha$ mRNA expression decreased gradually during the development of cardiac lesions in BIO14.6 (Katoh *et al.*, 1990). Thus, it appears that at the prenecrotic stage, the amount of G_s is within normal levels, but its functional activity starts to decline due to post-translational modification or other unknown cofactors. After the focal necrotic stage, expression of $G_s\alpha$ mRNA gradually decreases and, as a result, G_s is then decreased both quantitatively and qualitatively.

Although the precise roles of altered βAR and G_s in the pathogenesis of BIO14.6 have not been clear, recent results indicate the potential importance of the βAR–adenylyl cyclase system in maintaining intact myocardial structure and function in cardiomyopathy. Jasmin and Proschek (1983) observed that injection with the βAR agonist isoproterenol can reduce myofibrillar degeneration, myocyte Ca^{2+} content, and serum creatine kinase levels (a marker for muscle damage) in myopathic hamsters of another strain, UM-X7.1. Although the relevance of these data to BIO14.6 is only conjectural, the results suggest that reduced βAR–G_s activity and/or content might contribute to the development of hypertrophic cardiomyopathy. It is noteworthy that decreased G_s activity has also been observed in other animal models and in one clinical setting associated with cardiac hypertrophy: spontaneously hypertensive rat (SHR) (Murakami *et al.*, 1987), pressure-overloaded dog hearts (Longabaugh *et al.*, 1988), and transplanted human donor hearts (Dennis *et al.*, 1989a).

With regard to G_i, Katoh *et al.* (1990) reported a decrease in expression of G_i ($G_i2\alpha$) mRNA in hearts of BIO14.6 at the necrotic stage. On the other hand, Sen *et al.* (1990) reported an increase of pertussis toxin substrates in hearts of BIO14.6. Because pertussis toxin pretreatment completely interrupted the phenylephrine-induced increase of $[Ca^{2+}]_i$ in isolated myocytes of BIO14.6, the authors speculated that α_iAR may be coupled to the Ca^{2+} mobilization process through G_i in the heart of BIO14.6, and that increased G_i might contribute to the increased α_i response and Ca^{2+} overload in BIO14.6. Further investigations of G_i function and G_i mRNA, in particular of message stability, are needed.

Another animal model of cardiomyopathy has been reported in turkeys (Noren *et al.*, 1971). Basal, isoproterenol-stimulated, and NaF-stimulated adenylyl cyclase activities were significantly reduced in the heart in this model, and NaF-stimulated adenylyl cyclase activity was reduced even before development of cardiac dysfunction (Staley *et al.*, 1987). These indirect data suggest post-receptor alterations in the G-protein-linked adenylyl cyclase pathway in this model.

2.1.2 Idiopathic dilated cardiomyopathy (DCM) in humans

Because the term DCM is used to designate a condition in which heart muscle is damaged from unknown causes, the aetiology probably differs among patients. Alcohol abuse, hypertension, immunological disorders and viral infection are some of the causes of DCM. In some cases, hereditary predisposition is observed. In part, because of the diversity in the underlying cause of DCM, most experiments have focused on the alterations observed with congestive heart failure (CHF) in DCM patients. These results will be discussed in Section 2.2.2.

Recently, the existence of an anti-βAR autoantibody in DCM patients was reported by Limas *et al.* (1989). This autoantibody was able to inhibit the specific binding of the radioligand [³H]dihydroalprenolol to rat cardiac membrane βAR. Furthermore, the incidence of this anti-βAR autoantibody was closely related to the presence of the histocompatibility subtypes HLA-DR4 and HLA-DR1 (Limas *et al.*, 1990a, 1990b). It is not clear whether autoimmunity is directly related to the aetiology of disease in DCM patients, but conceivably infection or injury-induced damage to the heart could be a trigger for the production of anti-βAR antibodies, as suggested for autoantibodies that are detected in Chagas' disease (chronic *Trypansoma cruzi* infection) (Sterin-Borda *et al.*, 1986, 1988).

2.2 Congestive heart failure (CHF)

2.2.1 Animal models

Various experimental models of CHF have been used to study aspects of this pathological condition that cannot be easily assessed in human hearts (Smith and Nuttall, 1985). Among various models, we will discuss pressure overload, volume overload, and genetic cardiomyopathy.

Pressure overload Pressure overload-induced increase of βAR has been reported in several animals, including guinea pigs (Karliner *et al.*, 1980), rats (Limas, 1979), and dogs (Vatner *et al.*, 1985). Dog hearts, in which CHF was induced by pressure overload showed a 44% increase of βAR in ventricular myocardium (Vatner *et al.*, 1985), but the amount of G_s, determined by either cholera toxin-catalysed ADP-ribosylation or by reconstitution into S49 cyc⁻ cell membranes, was reduced 50–60% from control values (Longabaugh *et al.*, 1988). These results suggested that a deficiency of G_s could be a cause of reduced βAR responsiveness in this animal model. However, other changes were also observed in dogs with CHF. The number of muscarinic acetyl-choline (M_2) receptors was decreased by 36%, whereas the amount of G_i, as measured by pertussis toxin-catalysed ADP-ribosylation, was not changed. These data suggest that the impaired parasympathetic control observed in dogs with CHF is mainly due to the decreased M_2 receptor number (Vatner *et al.*, 1988a). The alteration of βAR in these pressure-overloaded animals contrasts with the observations in failing human hearts (Section 2.2.2). Thus, animal models of pressure overload might best be considered models for cardiac hypertrophy that can terminate in CHF. The mechanism(s) by which hypertrophic hearts develop heart failure in these animals may be different from that occurring in most cases of clinical CHF.

Pressure and volume overload Recently, a decrease of βAR was reported in failing rabbit hearts with combined pressure and volume overload-induced CHF (Gilson *et al.*, 1990). In this model, volume overload was produced by aortic insufficiency, and two weeks later pressure overload was added by aortic constriction. The number of βAR in the rabbits with CHF was reduced by 30% compared to control rabbits. Chamber-specific decrease of βAR was documented by Fan *et al.* (1987) in a dog right ventricular failure model created by pulmon-ary artery constriction (pressure overload) and tricuspid avulsion (volume overload). The density of βAR was decreased by 44% in right ventricle, but was not changed in left ventricle, despite an elevated plasma catecholamine concentration. Furthermore, both ventricles had a decreased positive inotropic response to dobutamine (β_1-adrener-gic agonist), and decreased isoproterenol-, Gpp(NH)p-, and forskolin-stimulated cyclase activity without a change in $MnCl_2$-stimulated cyclase activity. These observations suggested that the decrease of βAR was limited to the failing ventricle, but that the impaired contrac-tile response to βAR stimulation might be explained by G_s dysfunction.

It should be noted that sympathetic drive was presumably potentiated even in left ventricle, because myocardial catecholamine concentration was significantly reduced in both ventricles. In conjunction with an elevated circulating catecholamine concentration, which would act on both ventricles equally, there must be some difference between the two ventricles to explain the chamber-specific regulation of βAR. An impaired noradrenaline uptake activity in right ventricle has been demonstrated in this model, with significant correlation between the decrease in βAR density and that in noradrenaline uptake activity (Liang *et al.*, 1989). This study has suggested that both decreased noradrenaline uptake and increased sympathetic activity are needed to reduce βAR density in this model. Further studies of G-proteins in this model have not been reported but should be of interest. It is not yet fully understood whether variable findings regarding changes in βAR expression result from differences among animal species or experimental protocols used to produce heart failure. Measurement of noradrenaline concentration in the synaptic cleft, especially at an early stage of hypertrophy, may provide a clue for the mechanisms responsible for βAR regulation.

Genetic cardiomyopathy model The cardiomyopathic Syrian hamster strain BIO53.58 is characterized by progressive cardiac dilatation, and early death with CHF (Homburger, 1979). The most prominent difference from strain BIO14.6 is that animals in BIO53.58 do not develop cardiac hypertrophy before CHF. Because of this characteristic feature, BIO53.58 might be a more suitable animal model than BIO14.6 to study human DCM and CHF. Recently, the alterations in the βAR–adenylyl cyclase system of BIO53.58 have been reported at two different stages, 30 days (before CHF) and 100 days (CHF) (Feldman *et al.*, 1990). βAR density and affinity of BIO53.58 did not differ from those of control hamster (F1b) at either stage. The G_s activity of BIO53.58, as measured by reconstitution with S49 cyc$^-$ cell membranes, was significantly reduced at 100 days, though immunoblot analysis did not detect significant differences in the amount of immunoreactive $G_s\alpha$ at either age. This functional alteration of G_s in 100-day-old BIO53.58 was accompanied by a decrease of isoproterenol-stimulated adenylyl cyclase activity. In addition, reduced GTPγS- and forskolin-stimulated adenylyl cyclase activity has been found in 110- and 200-day-old BIO53.58, and the amount of G_i (assessed by pertussis toxin-catalysed ADP-ribosylation) was significantly increased in BIO53.58 at both ages (Urasawa *et al.*, 1990b). This increase in G_i was more prominent in

200-day-old BIO53.58. Furthermore, the increased G_i was functional, and could inhibit forskolin-stimulated adenylyl cyclase activity to a greater extent than in the control F1b animals (Urasawa, unpublished observation). These findings in GTP binding proteins of BIO53.58 are quite similar to those observed in failing human hearts (Section 2.2.2). Since these changes were not observed before development of heart failure in BIO53.58, and seemed to become prominent during the course of disease, the alterations in G-proteins would appear to be secondary phenomena to the cardiac dysfunction. The decreased G_s/G_i ratio may contribute to attenuated cardiac responsiveness to βAR stimulation in BIO53.58.

2.2.2 CHF in humans

Since various diseases and clinical settings can lead to CHF, it is unlikely that a single mechanism can explain the cardiac dysfunction observed in CHF patients. However, CHF patients share several similar features: reduced cardiac contractility, increased sympathetic nerve activity, and activation of many hormonal systems, such as the renin–angiotensin system, vasopressin, and atrial natriuretic factor (ANF), to compensate for the impaired cardiac function. In this setting, cardiac receptors, G-proteins and their effector molecules are exposed to chronically elevated neurohormonal stimulation. It is widely recognized that a chronic increase in such stimulation can desensitize target cell receptors and post-receptor signal transducing pathways (Clark, 1986). It is therefore likely that in CHF one or more types of desensitization contribute to a reduction in signal-transducing efficiency in order to maintain intracellular homeostasis. For example, decreased net activity of the βAR pathway has been reported by several laboratories: decreased positive inotropic response (Bristow *et al.*, 1982; Fowler *et al.*, 1986; Danielsen *et al.*, 1989; Harding *et al.*, 1990; Böhm *et al.*, 1988b, 1989b, 1990c); and decreased adenylyl cyclase response (Bristow *et al.*, 1982, 1989; Karliner and Scheinman, 1988; Böhm, 1989b; Mancini *et al.*, 1989). These changes in adrenergic responses are somewhat counterproductive because the failing heart largely depends on adrenergic neurohumoral support to overcome its impaired pump function.

Many laboratories have reported a decrease in βAR density in CHF patients (Bristow *et al.*, 1982; Dennis *et al.*, 1989a; Schwinger *et al.*, 1990). This is particularly prominent in terminally failing human hearts, and there is a close correlation between the extent of decrease in βAR density and severity of CHF (Bristow *et al.*, 1982; Böhm *et al.*, 1988a, 1988b). Because previous reports have indicated that $β_1AR$ and $β_2AR$ can be regulated in different ways (Hausdorff *et al.*, 1990), and that human heart contains both βAR subtypes (Jones *et al.*, 1989), receptor subtype-specific regulation in CHF may occur, but has not yet been well defined. Understanding of such regulation may prove important for directing clinical interventions. A $β_1AR$-selective decrease in failing hearts has been reported by several laboratories (Bristow *et al.*, 1989; Böhm *et al.*, 1989a). Although functional uncoupling of $β_2AR$ from post-receptor components has also been noted in such studies, preservation of $β_2AR$ density suggests that CHF patients might be treated with $β_2AR$-selective agonists in order to stimulate cardiac adenylyl cyclase and produce a positive inotropic effect in the failing human heart. In testing this hypothesis, Böhm and his colleagues found limited effects of the $β_2AR$-specific agonist, dopexamine, in producing inotropic support in patients with terminal CHF (Böhm *et al.*, 1990a). Alterations of βAR on CHF patients are discussed elsewhere (Brodde *et al.*, 1989; Bristow *et al.*, 1990; Feldman and Bristow, 1990; Horn and Bilezikian, 1990).

There are substantial amounts of data that indicate alterations of post-receptor components in CHF patients. Karliner and Scheinman (1988) reported possible post-receptor dysfunction in CHF patients, as based on decreased adenylyl cyclase activity in the presence of isoproterenol, Gpp(NH)p and NaF, in cardiac membranes prepared from biopsy samples of failing versus non-failing human hearts. Decreased Gpp(NH)p-stimulated cyclase activity in CHF patients has also been reported by Dennis *et al.* (1989a). Horn *et al.* (1988) reported that the amount of lymphocyte G_s, as measured by cholera toxin-catalysed ADP-ribosylation, was reduced by 80% in CHF patients, compared to values found in a control group. The authors speculated that decreased lymphocyte G_s might reflect alterations in cardiac tissue, because their other studies had revealed a close correlation between lymphocyte G_s and cardiac G_s. Moreover, the reductions of G_s and βAR density were reversible after successful treatment of CHF with angiotensin converting enzyme (ACE) inhibitors. Subsequent studies have not confirmed these conclusions of Horn *et al.* Maisel *et al.* (1990a) failed to find a substantial decrease in cholera toxin substrates in lymphocytes from patients with CHF, and two other groups reported no difference in the amount of ADP-ribosylated G_s between failing and non-failing human hearts (Feldman *et al.*, 1988; Schnabel *et al.*, 1990). Although data in these studies involve only relatively small sample numbers, the results suggest that quantitative analyses of G_s in human material in which toxin labelling is used may have problems with regard to labelling efficiency (Ransnas and Insel, 1988). Limited data have indicated an increased expression of $G_s\alpha$ mRNA in DCM patients with CHF (Feldman *et al.*, 1989a). Further investigations are needed to establish whether G_s function is changed, and whether such alterations modify cardiac contractility in failing human hearts.

Other studies have examined changes in G_i in CHF. Feldman *et al.* (1988) reported an increase of G_i in failing human hearts. Pertussis toxin-catalysed ADP-ribosylation revealed that the failing hearts of DCM patients had a 30% increase in G_i compared to values in non-failing hearts. At the same time, those workers found that there was no difference in the amount of G_s between failing and non-failing hearts, and that Gpp(NH)p-stimulated adenylyl cyclase activity was reduced in DCM patients in spite of the preserved forskolin-stimulated adenylyl cyclase activity. From these results, the authors speculated that increased amounts of G_i in patients with terminal CHF might act to tonically inhibit the G_s-activated adenylyl cyclase pathway. Feldman and colleagues have also shown increased expression of α_i3 mRNA in DCM hearts (Feldman *et al.*, 1989a). An increase in G_i in DCM patients has also been reported by other laboratories (Neumann *et al.*, 1988; Böhm *et al.*, 1989b). In the study by Böhm and his colleagues, the amount of G_i as assessed by pertussis toxin-catalysed ADP-ribosylation was increased by 37% compared to control, and this result was confirmed by immunoblot analysis using antiserum prepared against the C-terminal synthetic decapeptide of transducin. Böhm *et al.* (1990c) further reported that the increase of G_i in DCM hearts was associated with reduced basal and Gpp(NH)p-stimulated cyclase activity, and reduced responsiveness to isoprenaline. It should be noted that hearts from patients with ischaemic heart disease (IHD) did not show the increase of G_i, although they shared the same symptoms of CHF as did DCM patients. This suggests a role of increased G_i, perhaps most importantly in the pathogenesis of DCM. Recently, Maisel *et al.* (1990a) investigated the G_i function in lymphocytes of CHF patients (DCM and IHD) and reported that there was no significant difference in the amount of lymphocyte ADP-ribosylated G_i (and G_s) between CHF and control. However, pertussis toxin-induced potentiation of isoproterenol- or prostaglandin E_i-stimulated adenylyl cyclase activity was increased, suggesting potentiated G_i function in CHF patients. Moreover, the increase of cyclase activity was closely correlated with the plasma noradrenaline concentration.

Several laboratories have reported normal forskolin-stimulated cyclase activity in failing hearts (Fowler *et al.*, 1986; Böhm *et al.*, 1989b; Dennis *et al.*, 1989a,b). Relatively preserved responsiveness to histamine in failing hearts has also supported this concept (Bristow *et al.*, 1982, 1989). Recently, Harding *et al.* (1990) reported that single atrial myocytes isolated from NYHA class IV patients show reduced amplitude in maximal contractile response to forskolin. This would suggest that in terminally failing hearts the catalytic protein might be impaired, and could therefore contribute to post-receptor dysfunction.

Taken together, the data indicate that failing human hearts may have reduced βAR, reduced G_s function, increased G_i function, and perhaps decreased catalytic function. All these changes could contribute to a reduction of transmembrane signalling through βAR, or other receptors linked to stimulation of adenylyl cyclase (and perhaps Ca^{2+} channels). Alterations in the βAR–adenylyl cyclase system of failing hearts are summarized in Table 2.

2.3 Ischaemic heart disease (IHD)

Ischaemic heart disease is characterized by decreased oxygen supply to cardiac myocytes, and can be classified into two major categories. First is the acute myocardial infarction caused by coronary thrombosis, which in most cases leads to irreversible cell damage. Second is the angina pectoris which can be caused by episodic increases in oxygen demand and/or decreases in oxygen supply due to coronary spasm. In this case, myocardial ischaemia is reversible, and can be returned to a normal balanced condition by decrease in oxygen demand or cessation of coronary spasm if spasm has occurred.

It is well known that local sympathetic nerve activity can be markedly increased during ischaemia (Schömig *et al.*, 1984, 1987). Thus, ischaemia-induced alterations in the βAR-adenylyl cyclase system may contribute to mechanisms involved in myocardial cell damage, arrhythmia and heart failure that occur in this setting. For example, Corr *et al.* (1978) have found increased tissue cAMP concentration in ischaemic cat myocardium, and have speculated that increased adrenergic stimulation may contribute to the genesis of ventricular fibrillation. However, other data by Corr and co-workers have emphasized the importance of an increase in $α_1AR$ and signal transduction via phosphoinositide hydrolysis in experimental ischaemia (Heathers *et al.*, 1989).

Table 2. Changes in the βAR–G-protein–adenylyl cyclase system in congestive heart failure in humans

Reference	β	G_s	G_i	Adenylyl cyclase activity					
				B	I	G	F	FK	
Bristow et al. (1982)	↓→				→	▶[b]	↑		
Bristow et al. (1986)	↓[a]→			↑	▶		↑		
Brodde et al. (1986)	↓[a]→			↑	→			↑	
Fowler et al. (1986)	↓[a]→				→→		↑	↑	
Bristow et al. (1987)	↓[a]→				▶[c]		↑↑		
Böhm et al. (1988c)	↓[a]→								NYHA II,III vs IV
Bristow et al. (1988)	↓[a]→								
Feldman et al. (1988)	↓→	↑	↓	→		→	↑	↑	
Horn et al. (1988)	↓→	→		→					Lymphocyte of CHF patients
Karliner and Scheinman (1988)				→	▶[d]	→	→		
Näbauer et al. (1988)			↓	→	▶				
Neumann et al. (1988)		↑	↓↓		→→				
Böhm et al. (1989b)	↓[a]				→→				
Bristow et al. (1989)	↓→				→→	→	↑	↑	
Dennis et al. (1989b)	→					→		↑[e]	
Feldman et al. (1989a)	→	↑[f]	↑[f]						
Limas et al. (1989)	→→			→	→			↑	Mild CHF vs severe CHF
Mancini et al. (1989)	→→				→		↑		Lymphocyte of CHF patients

Danielsen et al. (1989)
Böhm et al. (1990c)
Harding et al. (1990)
Maisel et al. (1990a)

Schnabel et al. (1990)
Schwinger et al. (1990)

NYHA I,II vs III vs IV
Lymphocyte of CHF patients

▼[g]

↑ ▲
↑ ▲ ▲
↑[h]

↓

→[i]

→ ▲

↓

β, βAR density; B, basal cyclase activity; I, isoproterenol-stimulated cyclase activity; G, GTP analogue-stimulated cyclase activity; FK, forskolin-stimulated cyclase activity; F, NaF-stimulated cyclase activity; ▼, decreased physiological response; ▲, no change in physiological response.

[a] β_1AR ↓, β_2AR →
[b] Histamine ▲
[c] Dobutamine ▼, histamine ▼ (in NYHA IV)
[d] Histamine ▲, ouabain ▲
[e] Incubated with Mn^{2+}
[f] mRNA expression level
[g] Only in NYHA IV
[h] Assessed by pertussis toxin-induced potentiation
[i] Incubated with ADP-ribosylation factor (ARF)

An increase in βAR density in ischaemic myocardium was first reported by Mukherjee *et al.* (1979). These investigators showed that experimental myocardial ischaemia produced in dogs by coronary artery ligation was accompanied by a rapid increase in βAR density in association with a marked decrease in noradrenaline concentration in the ischaemic tissues. Maisel *et al.* (1985) demonstrated the ischaemia-induced increase of membrane-associated βAR using a guinea pig infarction model produced by coronary artery ligation. Those workers also prepared a light vesicle (presumably intracellular) fraction from ischaemic myocardium, and found that βAR density in this fraction was decreased by ischaemia. Since total βAR density of combined sarcolemmal and light vesicle fraction was constant after ischaemia, it was proposed that the origin of the increased sarcolemmal βAR might be the intracellular βAR pool. In addition, these workers observed an increase in isoproterenol-stimulated adenylyl cyclase activity in the membrane fractions of ischaemia tissue without a change in forskolin-stimulated adenylyl cyclase activity. Subsequent studies have not consistently replicated the increase in adenylyl cyclase activity (Maisel, personal communication).

Devos *et al.* (1985) found a 35% increase in βAR density, and significant decreases in basal, and isoprenaline-, Gpp(NH)p-, NaF- and forskolin-stimulated adenylyl cyclase activities in a canine model of acute ischaemia. In addition, the GTP-induced shift of β-adrenergic agonist competition curve was substantially reduced in ischaemic myocardium, with a reduction of high-affinity-state binding of agonist to βAR (43% in non-ischaemic to 20% in ischaemic). From these observations, the authors concluded that post-receptor components of the βAR–adenylyl cyclase system, especially G_s, may be altered in acute ischaemia.

Vatner *et al.* (1988b) have examined the βAR–adenylyl cyclase system of conscious dogs, in which myocardial ischaemia was produced by a hydraulic occluder surgically placed around the coronary artery. The dogs were allowed to recover from surgical stress for more than three weeks prior to ischaemia. Forty-five minutes of coronary occlusion increased βAR number about 50%, but decreased basal, and isoproterenol-, Gpp(NH)p-, NaF- and forskolin-stimulated adenylyl cyclase activity more than 40%. The attenuation in forskolin-stimulated cyclase activity in the presence of manganese suggested a dysfunction of catalytic protein itself. Using the same model, these workers have shown decreased G_s activity by reconstitution of S49 cyc^- membranes during ischaemia (Susanni *et al.*, 1989). Bovine hearts were treated using a similar protocol, and showed the same alterations in the βAR–adenylyl cyclase system, suggesting that the changes are not species-specific (Vatner *et al.*, 1990).

Recently, Strasser *et al.* (1990) used perfused rat hearts to assess ischaemia-induced changes in the βAR–adenylyl cyclase system. Global ischaemia of perfused rat hearts produced a rapid increase of βAR number, which was detectable within 15 min after the onset of ischaemia, but became more prominent over the subsequent 30 min. To test the hypothesis that energy depletion may contribute to the receptor upregulation, the authors perfused rat hearts with cyanide to reduce the production of high-energy phosphates. Perfusion with cyanide also rapidly increased βAR number to a level similar to that achieved with global ischaemia. Isoproterenol-stimulated adenylyl cyclase activity was increased after 15 min ischaemia, but decreased by 50 min to a lower level of activity than was found in control myocardium, and perfusion with cyanide could not reproduce these changes in adenylyl cyclase activity. Forskolin-stimulated adenylyl cyclase activity was also enhanced at 15 min but decreased by 50 min, and these changes were not affected by concomitant βAR antagonist treatment. These results suggest that the transient sensitization of the adenylyl cyclase system, which was observed after a short period of ischaemia, resulted from a modification of catalytic protein itself, and did not depend on βAR stimulation or intracellular energy depletion.

Wolff *et al.* (1989) have documented opposite results for the effect of ischaemia on βAR density. In their investigation, βAR density was measured using a 1000*g* pellet which is routinely discarded in many other investigations to eliminate the contamination of nucleus and contractile proteins, and this preparation showed more than a 50% reduction in receptor number compared to control. Moreover, they measured βAR density in the conventional membrane preparation (40 000*g* pellet), and showed that the 1000*g* pellet contained more than 70% of total βAR (sum of two fractions).

Limited data are available thus far regarding alterations in G-proteins in ischaemic heart disease. Some results have suggested that myocardial ischaemia can lead to upregulation of βAR, transient sensitization of catalytic protein, and decreased G_s activity, if conventional membrane preparations are used (Susanni *et al.*, 1989; Maisel *et al.*, 1990b). Although GTP-binding proteins may not have an active role in the pathogenesis of ischaemia-induced myocardial cell damage and arrhythmia, they might serve to modulate the cardiac function during myocardial ischaemia. Ischaemia-induced changes in the βAR–adenylyl cyclase system are summarized in Table 3.

Table 3. Changes in the βAR–G-protein–adenylyl cyclase system in ischaemic heart disease

Acute ischaemia

		Adenylyl cyclase activity						Species	Method
	β	G_s	B	I	G	F	FK		
Mukherjee et al. (1979)	↑							Dog	Coronary ligation (≤8 h)
Drummond and Sordahl (1981)									
Mukherjee et al. (1982)	↓		→	→				Dog	Coronary ligation (≤1 h)
Muntz et al. (1984)	↓							Dog	Coronary ligation (1 h)
Devos et al. (1985)	↓			→	→	→	→	Dog	Coronary ligation (1 h)
Maisel et al. (1985)	↑		→	↓[a]	↑	→	↑	Dog	Coronary ligation (5 h)
Will-Shahab et al. (1985)				→	→			Guinea pig	Coronary ligation (1.5 h)
Will-Shahab et al. (1986a,b)			↑	→		→		Rat	Anoxic incubation (≤1 h) Coronary ligation (20 min)
Maisel et al. (1987)	↓		→	←	→	→	↑	Rat	Coronary ligation (1 h)
Vatner et al. (1988b)	↓			→			→	Guinea pig	Coronary occlusion (45 min)
Karliner et al. (1989)	↑		↑	→	→	→		Dog^h	Coronary ligation (1–2 h)
Marsh and Sweeney (1989)	↓[a]	↓[b]		→		→		Dog	Hypoxic culture (2 h)
Susanni et al. (1989)	↓		↑	→	→	→	→	Chick embryo	Coronary occlusion (45 min)
Wolff et al. (1989)	↓[c]							Dog^h	Coronary ligation (30 min)
Strasser et al. (1990)	↓		↑	↓[d]	→	↑	[d]	Rabbit	Langendorff (≤50 min)
Vatner et al. (1990)	↓		→	→		→	→	Rat	Coronary occlusion (1 h)
								$Calf^h$	

Chronic ischaemia

	β	G_s	B	I	G	F	FK	Species	Method
Baumann et al. (1981)	↓		↑	↓		↑		Guinea pig	Coronary ligation (3–6 days)
Karliner et al. (1986b)	↓[e]		↑	↓	↓	→		Dog	Coronary ligation (3 weeks)
Bevilacqua et al. (1986)	↑[f]		→[g]					Human	Myocardial infarction
Kammerling et al. (1987)				→[g]				Dog	Coronary ligation (5–10 days)
Böhm et al. (1990b)	↓							Human	Ischaemic cardiomyopathy

β, βAR density; B, basal cyclase activity; I, isoproterenol-stimulated cyclase activity; G, GTP analogue-stimulated cyclase activity; F, NaF-stimulated cyclase activity; FK, forskolin-stimulated cyclase activity.

[a] Reversible
[b] Reconstruction with S49 cyc⁻
[c] Used 1000g pellet
[d] ↑ at 15 min, ↓ at 50 min
[e] β₁AR
[f] β₁AR ↑, β₂AR ↓
[g] Used apical denervated myocardium
[h] Conscious at experiment

2.4 Hypertension

Hypertension in humans is composed of two major categories: essential hypertension and secondary hypertension. The latter category comprises settings, such as renal diseases, hormonal disorders, etc., which lead to an increase in blood pressure. By contrast, aetiology of essential hypertension still remains to be elucidated, although it is widely recognized that blood pressure is determined and controlled by multiple factors, including neuronal input and hormonal regulation of a variety of cell types. To date, investigations dealing with alteration of GTP-binding proteins in hypertensive subjects have been quite limited, although some results have been reported in animal models.

2.4.1 Animal models of hypertension

The spontaneously hypertensive rat (SHR) has been a widely used animal model to study possible alterations of signal-transducing pathways in hypertension and its associated cardiac hypertrophy. Numerous physiological studies have suggested that the positive inotropic effect of catecholamines is less effective in SHR myocardium and vasculature than in the normotensive control, the Wistar Kyoto rat (WKY) (Saragoça and Tarazi, 1981a; Fujimoto et al., 1987; Asano et al., 1988a; Böhm et al., 1988d). For example, the relaxant response due to activation of several different adenylyl cyclase-linked receptors, including A_2-adenosine, H_2-histamine and D_1-dopamine receptors, are reportedly decreased in SHR femoral artery (Asano et al., 1988b). Such heterologous subsensitivity suggests dysfunction of a common post-receptor mechanism in G_s-coupled receptors. Since forskolin-stimulated relaxant response of SHR was equivalent to that of WKY, it has been proposed that G_s function is attenuated in SHR femoral artery.

Several laboratories have reported a decrease of βAR density in SHR (Will-Shahab et al., 1986a; Limas and Limas, 1987; Böhm et al., 1988d), although it has been proposed that this decrease may be a consequence rather than a cause of hypertension (Michel et al., 1990). Böhm et al. (1988d) indicated that the effect of GTP in shifting the competition curve for the agonist isoprenaline was similar in SHR and WKY, suggesting that βAR–G_s coupling remained intact in SHR. This conclusion is supported by binding data obtained using isolated myocytes of SHR and age-matched WKY rats (Grammas et al., 1989).

By contrast, reduced G_s function in heart from SHR was reported by Murakami et al. (1987), who conducted reconstitution assays with brain adenylyl cyclase and detergent extracts from heart. These functional data differ from their other results in which the amount of G_s, as determined by cholera toxin-catalysed ADP-ribosylation, was similar in SHR and WKY. In other work, Anand-Srivastava (1988) has demonstrated an attenuation in GTP analogue-stimulated adenylyl cyclase activity in SHR, a result suggesting impaired G_s or catalyst activity as the mechanism of impaired βAR response in SHR.

Conflicting results have been reported for the alterations in the catalytic protein of adenylyl cyclase of SHR heart. Anand-Srivastava *et al.* (1987) reported decreased forskolin-stimulated adenylyl cyclase activity, suggesting impaired catalytic function. Since it is now widely recognized that the maximal response to forskolin is affected by the degree of G_s activation (Darfler *et al.*, 1982; Wong and Martin, 1983; Seamon and Daly, 1986), it is dubious whether catalytic activity apart from G_s function can be measured by forskolin-stimulated adenylyl cyclase activity. In fact, Murakami *et al.* (1987) have demonstrated an increase in catalytic activity of SHR heart using purified G_s as an exogenous stimulator.

Intracellular calcium has been reported to be elevated in various tissue and blood cells of SHR (Zidek *et al.*, 1982; Bruschi *et al.*, 1985; Sugiyama *et al.*, 1986). This has suggested abnormalities in calcium handling, and, as a result, the phosphoinositide turnover pathway has been investigated by several laboratories. An increased phospholipase C (PLC) activity in aorta of prehypertensive SHR was reported by Uehara *et al.* (1988). Millanvoye *et al.* (1988) have shown that isolated aortic smooth muscle cells from SHR have an increased angiotensin II-induced PLC activity compared to cells from WKY. Contrasting data have been reported by Jeffries *et al.* (1988), who observed decreased agonist-promoted phosphoinositide hydrolysis in kidney of SHR. Recently, Makita and Yasuda (1990) have reported that membrane-bound phosphatidylinositol 4,5-bisphosphate hydrolysing phospholipase C (PIP_2-PLC) activity is increased in the hearts of SHR. Although the G-protein that couples receptors to PLC in heart tissue has not yet been identified, it may be worthwhile to investigate whether changes in newly recognized G-proteins, such as G_q, occur in the cardiovascular system of SHR and other hypertensive strains.

Acquired models of hypertension may also show abnormalities in receptors and post-receptor components (Michel *et al.*, 1990). For example, cardiac hypertrophy in renal hypertensive rat (RHR) is associated with reduced contractile response to β-adrenergic stimulation (Saragoça and Tarazi, 1981b). Using Langendorff-perfused hearts, Fouad et al. (1986) showed that response to glucagon and vasoactive intestinal peptide (VIP), which are coupled to adenylyl cyclase through G_s, was significantly reduced in RHR compared with sham-operated normotensive rats (heterologous desensitization). Other preliminary data have indicated that the expression of $G_s\alpha$ and $G_i2\alpha$ mRNA is suppressed in DOCA/salt hypertension rats (Sobierai et al., 1989). These results provide further evidence that G-proteins and/ or the catalytic protein are involved in the dysfunction in this animal model.

Since kidney cross-transplantation can show dramatic effects on blood pressure in many animal models (Dahl et al., 1974; Bianchi et al., 1974; Dahl and Heine, 1975; Kawabe et al., 1979), altered renal function may be directly related to the aetiology of certain forms of hypertension, and alterations in the βAR–adenylyl cyclase system of hypertensive animals might be secondary phenomena to the increased workload, increased sympathetic nerve activity, and changes in humoral conditions. Among various animal models of hypertension, such as SHR, RHR, Dahl S rats with high sodium diet, and DOCA/salt hypertensive rats only SHR had an increased renal αAR density, while an increase of $\beta_1 AR$ and $\beta_2 AR$ densities in kidney was observed in all hypertensive models (Michel et al., 1989). Since renal αAR can act on tubular sodium reabsorption, the authors speculated that the increased αAR activity in kidney may be a key phenomenon in the development of genetic hypertension. Clearly, more data are required to test this hypothesis.

2.4.2 Hypertension in humans

Substantial data have been published implicating possible changes in AR in the pathogenesis of hypertension, as reviewed elsewhere (Michel et al., 1990). To date, no direct studies of G-protein have been reported in hypertensive patients; only indirect evidence is available regarding these proteins. For example, a decrease in βAR-mediated adenylyl cyclase activity was reported in lymphocytes of hypertensive patients (Feldman et al., 1987). This decrease was accompanied by a parallel reduction of agonist affinity at βAR without decrease in βAR density. Thus, functional uncoupling of βAR from G_s was suggested. Although the precise mechanism of these changes has not been defined, it is noteworthy that a low-sodium diet could correct these alterations in the lymphocyte βAR pathway (Feldman et al., 1987; Feldman, 1987).

Other evidence also suggests that post-receptor components may be altered in hypertension. Thus, Schultz *et al.* (1989) examined lymphocyte adenylyl cyclase activity in normotensive volunteers with and without a positive family history of hypertension. These workers found that forskolin-stimulated adenylyl cyclase activity was significantly higher in the group with a positive family history. It was thus proposed that forskolin-stimulated adenylyl cyclase activity in lymphocytes might be related to as yet unknown hereditary factors associated with familial predisposition to essential hypertension.

3 SIGNALLING ABNORMALITIES INDUCED BY DRUG TREATMENT

Various chemical substances are likely to alter the receptor pathway linked to adenylyl cyclase. Theoretically, drugs that increase intracellular cAMP concentration, such as agonists for G_s-coupled receptors (catecholamines, glucagon, histamine, VIP, etc.) and cAMP phosphodiesterase inhibitors, might produce heterologous desensitization. Agents that act on the sympathetic or parasympathetic nerve systems might also alter the signal-transducing efficiency of the cardiac βAR–adenylyl cyclase system by modulating signal input to the system. Additionally, thyroid hormone, steroid hormones and cytokines may alter the transcription rate of one or more genes that mediate signal transduction. However, many of these possibilities remain to be investigated. We will describe recent data and general concepts regarding catecholamine-induced alterations in the cardiac βAR-adenylyl cyclase system.

3.1 Catecholamines

Noradrenaline, isoproterenol, dopamine, and dobutamine are catecholamines that are used clinically to maintain blood pressure, cardiac output, and renal blood flow, especially in patients with severe CHF. However, in most cases, chronic administration of these agents can yield gradually declining responsiveness of the cardiovascular system. Such phenomena, termed 'desensitization', have been intensively studied in various tissues and cell lines (Bouvier, 1990; Lefkowitz *et al.*, 1990).

A catecholamine-induced decrease of βAR number and adenylyl cyclase activity in the cardiovascular system was first reported by Tse *et al.* (1979). The number of cardiac βAR was reduced to 60% of the control value in rats injected with isoproterenol for 10 days. The changes in βAR were accompanied by a decrease in high-affinity binding of agonist to βAR, isoproterenol-stimulated adenylyl cyclase activity, and PKA activity. Although information regarding structural details of the βAR–adenylyl cyclase system was not available at the time this study was conducted, the authors speculated about involvement of non-receptor components of the adenylyl cyclase system in these alterations. They also demonstrated that the isoproterenol-induced changes were reversible after stopping isoproterenol injection.

Isoproterenol-induced heterologous desensitization in the heart was reported by Chatelain *et al.* (1982). After a five-day injection of isoproterenol, rat heart membranes showed impaired responsiveness to glucagon and serotonin as well as to isoproterenol. Chang *et al.* (1982) implanted osmotic mini-pumps in rats to obtain a constant rate of injection of catecholamines (isoproterenol or noradrenaline). Isoproterenol-induced positive inotropic and chronotropic effects were reduced as early as two hours after the implantation without any changes in βAR density. This result suggested that mechanisms other than a decrease in βAR number (downregulation) were operating at an early stage of the process, a phenomenon now commonly termed 'uncoupling'.

Rather different data have been reported by Vatner *et al.* (1989). In their study, dogs were injected with noradrenaline using osmotic mini-pumps for four weeks. Contrary to the earlier reports, βAR density in myocardium increased by 50% after noradrenaline infusion. However, basal, and isoproterenol-, Gpp(NH)p-, NaF-, and forskolin-stimulated adenylyl cyclase activities, and high affinity binding sites, were reduced significantly, suggesting dysfunction of the coupling mechanism between increased βAR and effector molecules. The authors speculated that an intact baroreceptor reflex and noradrenaline uptake mechanism might act to reduce noradrenaline concentration within the synaptic cleft, and thereby might cause increased βAR density. Moreover, these workers quantitated G_s by reconstitution in S49 cyc⁻ cell membranes and found a decrease in G_s activity in the noradrenaline-infused dogs. Conversely, Sato (1990) reported a noradrenaline-induced decrease of βAR density in rats. Two weeks infusion of subpressor doses of noradrenaline with osmotic mini-pumps had induced a significant reduction in βAR and cyclase activity of rat hearts. Their results indicate that animal species may affect the direction of βAR regulation.

So far, many laboratories have reported a catecholamine-induced decrease of βAR and impaired responsiveness of adenylyl cyclase (summarized in Table 4). However, information regarding the effect of long-term treatment of catecholamine on G-proteins is limited. Vatner *et al.* (1989) reported a decreased amount and activity of G_s and this presumably contributes to the impaired catecholamine responsiveness observed in noradrenaline-infused dogs. Reithmann *et al.* (1989) proposed another possibility to explain the catecholamine subsensitivity. Cultured neonatal rat heart cells exposed to $1 \mu M$ noradrenaline showed impaired responsiveness of adenylyl cyclase (similar to what has been observed in *in vivo* models), and also increased G_i, as assessed by pertussis toxin-catalysed ADP-ribosylation and Western blot analysis. The authors speculated that an increase in the level of G_i might contribute to the decreased sensitivity to catecholamines observed in various *in vivo* and *in vitro* systems. In fact, an increase of G_i was recently reported in hearts from noradrenaline-infused rats (Sato, 1990).

In summary, the alterations observed in chronic catecholamine treatment generally involve a decrease in βAR density, although other post-receptor changes seem likely. It is possible that decreased G_s activity, increased G_i, and decreased catalytic activity may also occur in the desensitized state (see Chapter 1 for further discussion). All these changes are quite similar to those observed in failing human hearts, thus suggesting that alterations in the βAR–adenylyl cyclase system of failing hearts are a consequence of the chronic catecholamine stimulation that characterizes heart failure. Catecholamine-induced alterations in the βAR–adenylyl cyclase system are summarized in Table 4.

3.2 Other drugs and biochemical substances

3.2.1 *β-Adrenergic receptor antagonists*

There have been several reports suggesting that abrupt withdrawal of propranolol can cause severe symptoms in patients with angina pectoris (Slome, 1973; Diaz *et al.*, 1973; Olson *et al.*, 1975). In some cases, abrupt cessation of propranolol can lead to an increased frequency of angina pectoris, palpitation, arrhythmia, and even acute myocardial infarction. This phenomenon has been called the 'propranolol (or β-

Table 4. Catecholamine effects on the adenylyl cyclase system

In vivo studies

	Adenylyl cyclase activity								Species	Catecholamine
	β	G_s	G_i	B	I	G	F	FK		
Tse *et al.* (1979)	→				→		→		Rat	Isoproterenol (3 mg/kg, 2/day, s.c., 10 days)
Chang *et al.* (1982)	→				▼				Rat	Isoproterenol (0.4 mg/kg per h, s.c., ≤7 days)
Chatelain *et al.* (1982)	→				↓[a]	→	→		Rat	Isoproterenol (0.25–5 mg/kg, 3/day, i.p., −5 days)
Maisel *et al.* (1988)	→				→	↑		↑	Guinea pig	Isoproterenol (0.05–0.2 mg/kg, s.c., one shot) Isoproterenol (0.15 mg/kg per h, s.c., ≤21 days)
Nanoff *et al.* (1989)	↓[b]			→	→				Rat	Isoproterenol (0.4 mg/kg per h, s.c., ≤7 days)
Vatner *et al.* (1989)	↑	↓[c]		→	→	→	→	→	Dog	Noradrenaline (0.03 mg/kg per h, s.c., 3–4 weeks)
Sato (1990)	→	↑			→	→	→	→	Rat	Noradrenaline (0.015 mg/kg per h, s.c., 14 days)

	β	G_s	G_i	B	I	G	F	FK	Tissue source[d]	Catecholamine
Bovik et al. (1981)	↓[e]				↓		↑		Embryonic chick HC	Isoproterenol (−100 μM, ≤16 h)
Marsh et al. (1982)					▼				Embryonic chick HC	Isoproterenol (1 μM, ≤3 h)
Limas and Limas (1984)	↓[f]				↓				Adult rat HC	Isoproterenol (1 μM, ≤20 min)
Karliner and Simpson (1988)	↓			↑	↓	↑	↑		Neonatal rat HC	Noradrenaline (1 μM, 2 days)
Reithmann and Werdan (1988)	↓				↓			→	Embryonic chick HC	Noradrenaline (1 μM, 3 days)
Reithmann et al. (1989)			↑[g]		↓	→	↑	→	Neonatal rat HC	Noradrenaline (1 μM, 3 days)
Reithmann and Werdan (1989)	↓			→	↓			→	Neonatal rat HC	Noradrenaline (1 μM, −5 days)

β, βAR density; B, basal cyclase activity; I, isoproterenol-stimulated cyclase activity; G, GTP analogue-stimulated cyclase activity; F, NaF-stimulated cyclase activity; FK, forskolin-stimulated cyclase activity; ▼, decreased physiological response.

[a] Glucagon ↓, secretin ↓
[b] $β_2AR$ ↓↓
[c] Determined by reconstitution assay and cholera toxin labelling
[d] HC = heart cells
[e] βAR density recovered within 24 h
[f] βAR density recovered within 20 min
[g] Determined by Western analysis and pertussis toxin labelling

blocker) withdrawal syndrome'. Several laboratories have shown cate-cholamine supersensitivity in animals or people after withdrawal of propranolol (Boudoulas *et al.*, 1977; Manning *et al.*, 1981; Webb *et al.*, 1981; Kennedy and Donnelly, 1982; Tenner, 1983a,b; Chess-Williams and Broadley, 1984; Cramb *et al.*, 1984). However, other studies have failed to show such supersensitivity (Myers and Horwitz, 1978; Myers *et al.*, 1979; Lindenfeld *et al.*, 1980). Conflicting results have also been shown in biochemical studies. Increased βAR density after propranolol treatment was reported by Glaubiger and Lefkowitz (1977), Aarons *et al.* (1980), Aarons and Molinoff (1982), Cramb *et al.* (1984) and Maisel *et al.* (1986), while other investigators showed no difference in βAR density after propranolol treatment (Baker and Potter, 1980; Kennedy and Donnelly, 1982; Chess-Williams and Broadley, 1984; Mügge *et al.*, 1985; Karliner *et al.*, 1986a). The reason for these discrepancies might be differences in the dose of propranolol, duration of treatment, and experimental animals used in the experiments. Direct evidence of effects of βAR antagonists on post-receptor events is not yet available. However, limited evidence that chronic treatment with βAR antagon-ists can enhance βAR–G_s coupling in the cardiovascular system (Cooper *et al.*, 1986; Hall *et al.*, 1990) suggests that post-receptor changes may occur in the setting of antagonist therapy. This type of change, enhanced βAR–G_s coupling, has also been reported in patients with mitral valve prolapse, a common cardiac valvular disorder in humans. A subset of the patients with mitral valve prolapse show increased responsiveness to isoproterenol with an increased percent-age of high-affinity state βAR, but without change in the number of βAR in neutrophils, a finding suggesting enhanced coupling of recep-tor to G_s (Davies *et al.*, 1987). The effects of chronic β-blocker treat-ment on the βAR–adenylyl cyclase system have been reviewed recently (Karliner, 1989; Becker *et al.*, 1988).

3.2.2 Thyroid hormone

Patients with hyperthyroidism show several clinical characteristics which are quite similar to those observed in hyperadrenergic states (Wildenthal, 1972; Roncari and Murthy, 1975). Much previous work has focused on possible increases in βAR as contributing to clinical manifestations of hyperthyroidism (Ciaraldi and Marinetti, 1978; Tse *et al.*, 1980; Stiles and Lefkowitz, 1981; Krawietz *et al.*, 1982; Hammond *et al.*, 1987; Hohl *et al.*, 1989). Changes in G-proteins in hyperthyroid and hypothyroid states were recently investigated by Levine *et al.*

(1990). The hyperthyroid state did not alter either protein or mRNA level of $G_s\alpha$, $G_i2\alpha$, $G_i3\alpha$, and G-protein β subunit ($G\beta$) in rat heart. By contrast, hypothyroid rats that had been treated with propylthiouracil showed a significant increase in both protein and mRNA level of $G_s\alpha$, $G_i2\alpha$, $G_i3\alpha$ and $G\beta$ mRNA, although the authors failed to find increased G_i function (guanine nucleotide-induced inhibition of forskolin-stimulated adenylyl cyclase activity). Previous work by Malbon *et al.* (1985) indicated that hypothyroidism may be associated with increased G_i in rat adipocytes. Since the effect of thyroid hormone on the βAR–G-protein–adenylyl cyclase system seems to be specific for each tissue, and perhaps for different species (Malbon *et al.*, 1988), further studies are needed to clarify the role of post-receptor components in the cardiovascular system in different thyroid states. The effects of thyroid state on the βAR–adenylyl cyclase system have been reviewed elsewhere (Bilezikian and Loeb, 1983; Stiles *et al.*, 1984; Malbon *et al.*, 1988) (see Chapter 7 for further discussion).

3.2.3 Miscellaneous

G-proteins have been suggested to be a target of other therapeutic agents. Lee *et al.* (1989) have reported that interleukin 1β (IL-1β, a chemical mediator active in inflammation) can increase $G_i2\alpha$ mRNA in vascular endothelial cells without changing mRNA level of G_i2, G_i3 or G_o, suggesting that expression of members of the G_i family are regulated separately. *In vitro* studies have indicated that ethanol can induce an enhancement of G_s–catalytic protein coupling in rabbit hearts (Feldman *et al.*, 1989b), and halothane can interrupt muscarinic (M_2) receptor–G_i coupling in rat hearts (Narayanan *et al.*, 1988). Chronic administration of doxorubicin (a cancer chemotherapeutic agent) in mice can potentiate cardiac adenylyl cyclase activity stimulated by GTP, isoproterenol, NaF, and forskolin (Robinson and Giri, 1986), although the mechanisms of these changes have not been identified.

4 OVERVIEW AND FUTURE PROSPECTS

As we have described, a rapidly expanding body of work has accrued in which changes in receptors (in particular, β-adrenergic receptors) and signal transduction by those receptors has been examined in humans and animals with cardiovascular disorders. It is difficult at this point to draw exact conclusions, but certain patterns seem to be emerging. Thus, changes in heart failure may in part depend on the aetiology of the disease, but, as heart failure progresses, the changes may, in large part, be secondary to catecholamines or perhaps other neuronally derived factors. Changes in myocardial ischaemia may involve the loss in energy-dependent regulation of receptors and post-receptor components, and perhaps covalent modification of the catalyst of adenylyl cyclase. In hypertension, changes in receptors and/or G-proteins may be involved in some aspects of the disease, but it has not been possible to link conclusively such changes to pathogenesis. Thus, at this time, no definitive evidence has been provided that receptors and/or post-receptor components are of primary patho-genetic importance in major cardiovascular disorders. This somewhat 'pessimistic' conclusion does not, however, mean that changes in receptors and post-receptor components are not of importance in contributing to progression of disease and clinical manifestation of the disorders.

The application of a variety of new tools is likely to prove critical for clarifying the role of post-receptor components in cardiovascular disorders. Examples include cDNA probes for quantitation of G-protein subunits in heart and vessels, perhaps involving use of PCR and *in situ* hybridization, use of specific antibodies in ELISA or quantitative Western blotting, and methods to more directly charac-terize changes in effector molecules. It is not yet possible to define whether covalent modifications in G-protein or effector molecules occur in disease states. Methods that involve microsequencing and chemical assessment of such changes will probably be needed.

There are several further biological aspects of G-proteins whose clinical importance remains to be determined. Firstly, although cer-tain $G\alpha$ subunits can link to multiple effectors, the possibility that such linkages may change in disease has not been explored. Secondly, a growing body of work suggests that the $G\alpha$ subunits, in particular α_s, may exist in compartments other than the plasma membrane (Roth *et al.*, 1991). Does such compartmentation change in disease? Thirdly, increasing evidence points to 'cross-talk' between signal-transducing pathways (e.g. effects of G_q-linked phosphoinositide hydrolysis and its resultant inositol phosphates and diacylglycerol on signal transduc-tion mediated by G_s or G_i). Disease-related alterations in such 'cross-talk' remain largely unexplored.

In writing this chapter, we have been struck by the rapid growth in number of recent publications in this field and the burgeoning interest that the topic has generated. We sense that we have tried to take a snapshot of a rapidly moving train. Given the clinical importance of the disorders that we have discussed, the application of more sophisticated methodology to well-selected patient groups and animal models would seem likely to yield new and, we would hope, clinically useful insights in the next several years.

ACKNOWLEDGEMENTS

Work in the authors' laboratory is supported by grants from NIH and post-doctoral fellowships (to K.U.) from the American Heart Association, California Affiliate, and the Japan Heart Foundation.

REFERENCES

Aarons, R.D. and Molinoff, P.B. (1982). *J. Pharmacol. Exp. Ther.* **221**, 439–443.

Aarons, R.D., Nies, A.S., Gal, J., Hegstrand, L.R. and Molinoff, P.B. (1980). *J. Clin. Invest.* **65**, 949–957.

Alexander, R.W., Brock, T.A., Gimbrone, M.A. Jr and Rittenhouse, S.E. (1985). *Hypertension* **7**, 447–451.

Alousi, A., Jasper, J.R., Insel, P.A. and Motulsky, H.J. (1991). *FASEB J.* **5**, 2300–2303.

Anand-Srivastava, M.B. (1988). *Biochem. Pharmacol.* **37**, 3017–3022.

Anand-Srivastava, M.B., Srivastava, A.K. and Cantin, M. (1987). *J. Biol. Chem.* **262**, 4931–4934.

Asano, M., Masuzawa, K., Matsuda, T. and Asano, T. (1988a). *J. Pharmacol. Exp. Ther.* **246**, 709–718.

Asano, M., Masuzawa, K. and Matsuda, T. (1988b). *Br. J. Pharmacol.* **95**, 241–251.

Bajusz, E., Homburger, F., Baker, J.R. and Opie, L.H. (1966). *Ann. N. Y. Acad. Sci.* **138**, 213–231.

Baker, S.P. and Potter, L.T. (1980). *Br. J. Pharmacol.* **68**, 8–10.

Barber, R. (1988). *Second Messengers and Phosphoproteins* **12**, 59–71.

Baumann, G., Riess, G. Erhardt, W.D., Felix, S.B., Ludwig, L., Blümel, G. and Blömer, H. (1981). *Am. Heart J.* **101**, 569–581.

Becker, D.V., Romano, F.D., Scanlon, P.J., Jones, S.B. and Euler, D.E. (1988). *Pharmacology* **36**, 172–182.

Bevilacqua, M., Norbiato, G., Vago, T., Meroni, R., Dagani, R., Raggi, U., Frigeni, G. and Santoli, C. (1986). *Eur. J. Clin. Invest.* **16**, 163–168.

Bianchi, G., Fox, U., Di Francesco, G.F., Giovanetti, A.M. and Pagetti, D. (1974). Clin. Sci. Mol. Med. **47**, 435–448.

Bilezikian, J.P. and Loeb, J.N. (1983). Endocrine Rev. **4**, 378–388.

Birnbaumer, L., Abramowitz, J., Yatani, A., Okabe, K., Mattera, R., Graf, R., Snaford, J., Codina, J. and Brown, A.M. (1990). Biochem. Mol. Biol. **25**, 225–244.

Böhm, M., Beuckelmann, D., Brown, L., Feiler, G., Lorenz, B., Näbauer, M., Kemkes, B. and Erdmann, E. (1988a). Eur. Heart J. **9**, 844–852.

Böhm, M., Diet, F., Feiler, G., Kemkes, B., Kreuzer, E., Weinhold, C. and Erdmann, E. (1988b). J. Cardiovasc. Pharmacol. **12**, 726–732.

Böhm, M., Diet, F., Feiler, G., Kemkes, B. and Erdmann, E. (1988c). J. Cardiovasc. Pharmacol. **12**, 357–364.

Böhm, M., Beuckelmann, D., Diet, F., Feiler, G., Lohse, M.J. and Erdmann, E. (1988d). Naunyn-Schmiedeberg's Arch. Pharmacol. **338**, 383–391.

Böhm, M., Pieske, B., Schnabel, P., Schwinger, R., Kemkes, B., Klövekorn, W.-P. and Erdmann, E. (1989a). J. Cardiovasc. Pharmacol. **14**, 549–559.

Böhm, M., Gierschik, P., Jakobs, K.H., Schnabel, P., Kemkes, B. and Erdmann, E. (1989b). Am. J. Cardiol. **64**, 812–814.

Böhm, M., Reuschel-Janetschek, E. and Erdmann, E. (1990a). Am. J. Cardiol. **65**, 395–396.

Böhm, M., Ungerer M. and Erdmann, E. (1990b). Am. J. Cardiol. **66**, 880–882.

Böhm, M., Gierschik, P., Jakobs, K.H., Pieske, B., Schnabel, P., Ungerer, M. and Erdmann, E. (1990c). Circulation **82**, 1249–1265.

Boudoulas, H., Lewis, R.P., Kates, R.E. and Dalamangas, G. (1977). Ann. Intern. Med. **87**, 433–436.

Bouvier, M. (1990). Ann. N. Y. Acad. Sci. **594**, 120–129.

Bocik, A., Campbell, J.H., Carson, V. and Campbell, G.R. (1981). J. Cardiovasc. Pharmacol. **3**, 541–553.

Bristow, M.R., Ginsburg, R., Minobe, W., Cubicciotti, R.S., Sageman, W.S., Lurie, K., Billingham, M.E., Harrison, D.C. and Stinson, E.B. (1982). N. Engl. J. Med. **307**, 205–211.

Bristow, M.R., Ginsburg, R., Umans, V., Fowler, M., Minobe, W., Rasmussen, R., Zera, P., Menlove, R., Shah, P., Jamieson, S. and Stinson, E.B. (1986). Circ. Res. **59**, 297–309.

Bristow, M.R., Ginsburg, R., Gilbert, E.M. and Hershberger, R. E. (1987). Basic Res. Cardiol. **82** (suppl. 2), 369–376.

Bristow, M.R., Minobe, W., Rasmussen, R., Hershberger, R.E. and Hoffman, B.B. (1988). J. Pharmacol. Exp. Ther. **247**, 1039–1045.

Bristow, M.R., Hershberger, R.E., Port, J.D., Minobe, W. and Rasmussen, R. (1989). Mol. Pharmacol. **35**, 295–303.

Bristow, M.R., Hershberger, R.E., Port, J.D., Gilbert, E.M. Snadoval, A., Rasmussen, R., Cates, A.E. and Feldman, A.M. (1990). Circulation **82** (suppl. I), I-12–I-25.

Brodde, O.-E., Schemuth, R., Brinkmann, M., Wang, X.L., Daul, A. and Borchard, U. (1986). Naunyn-Schmiedeberg's Arch. Pharmacol. **333**, 130–138.

Brodde, O.-E., Michel, M.C., Gordon, E.P., Sandoval, A., Gilbert, E.M. and Bristow, M.R. (1989). *Eur. Heart J.* **10** (suppl. B), 2–10.

Brown, J.H., Buxton, I.L. and Brunton, L.L. (1985). *Circ. Res.* **57**, 532–537.

Bruns, C. and Marmé, D. (1987). *FEBS Lett.* **212**, 40–44.

Bruschi, G., Bruschi, M.E., Caroppo, M., Orlandini, G., Spaggiari, M. and Cavatorta, A. (1985). *Clin. Sci.* **68**, 179–184.

Chang, H.Y., Klein, R.M. and Kunos, G. (1982). *J. Pharmacol. Exp. Ther.* **221**, 784–789.

Chatelain, P., Robberecht, P., De Neef, P., Descodt-Lanckman, M., Konig, W. and Christophe, J. (1980). *Pflugers Arch.* **389**, 21–27.

Chatelain, P., Robberecht, P., De Neef, P., Camus, J.C. and Christophe, J. (1982). *Biochem. Pharmacol.* **31**, 347–352.

Chess-Williams, R.G. and Broadley, K.J. (1984). *J. Cardiovasc. Pharmacol.* **6**, 701–706.

Ciaraldi, T.P. and Marinetti, G.V. (1978). *Biochim. Biophys. Acta.* **541**, 334–346.

Clark, R.B. (1986). *Adv. Cyclic Nucleotide Protein Phosphorylation Res.* **20**, 151–209.

Cooper, G., Kant, R.L., McGonigle, P. and Watanabe, A.M. (1986). *J. Clin. Invest.* **77**, 441–445.

Corr, P.B., Witkowski, F.X. and Sobel, B.E. (1978). *J. Clin. Invest.* **61**, 109–119.

Cramb, G., Griffiths, N.M., Aiton, J.F. and Simmons, N.L. (1984). *Biochem. Pharmacol.* **33**, 1969–1976.

Dahl, L.K. and Heine, M. (1975). *Circ. Res.* **36**, 692–696.

Dahl, L.K., Heine, M. and Thompson, K. (1974). *Circ. Res.* **34**, 94–101.

Danielsen, W., v. der Leyen, H., Meyer, W., Neumann, J., Schmitz, W., Scholz, H., Starbatty, J., Stein, B., Doring, V. and Kalmer, P. (1989). *J. Cardiovasc. Pharmacol.* **14**, 171–173.

Darfler, F.K., Mahan, L.C., Koachman, A.M. and Insel, P.A. (1982). *J. Biol. Chem.* **257**, 11901–11907.

Davies, A.O., Mares, A., Pool, J.L. and Taylor, A.A. (1987). *Am. J. Med.* **82**, 193–201.

Dennis, A.R., Marsh, J.D., Quigg, R.J., Gordon, J.B. and Colucci, W.S. (1989a). *Circulation* **79**, 1028–1034.

Dennis, A.R., Colucci, W.S., Allen, P.D. and Marsh, J.D. (1989b). *J. Mol. Cell. Cardiol.* **21**, 651–660.

Devos, C., Robberecht, P., Nokin, P., Waelbroeck, M., Clinet, M., Camus, J.C., Beaufort, P., Schoenfeld, P. and Christophe, J. (1985). *Naunyn-Schmiedeberg's Arch. Pharmacol.* **331**, 71–75.

Diaz, R.G., Somberg, J.C., Freeman, E. and Levitt, B. (1973). *Lancet* **1**, 1068.

Drummond, R.W. and Sordahl, L.A. (1981). *J. Mol. Cell. Cardiol.* **13**, 323–330.

Fan, T.-H.M., Liang, C., Kawashima, S. and Banerjee, S.P. (1987). *Eur. J. Pharmacol.* **140**, 123–132.

Feldman, R.D. (1987). *Can. J. Physiol. Pharmacol.* **65**, 1666–1672.

Feldman, A.M. and Bristow, M.R. (1990). *Cardiology* **77** (suppl. 1), 1–32.

Feldman, R.D., Lawton, W.J. and McArdle, W.L. (1987). *J. Clin. Invest.* **79**, 290–294.

Feldman, A.M., Cates, A.E., Veazey, W.B., Hershberger, R.E., Bristow, M.R., Baughman, K.L., Baumgartner, W.A. and Van Dop, C. (1988). *J. Clin. Invest.* **82**, 189–197.

Feldman, A.M., Cates, A.E., Bristow, M.R., and Van Dop, C. (1989a). *J. Mol. Cell Cardiol.* **21**, 359–365.

Feldman, A.M., Levine, M.A., Cates, A.E., Baughman, K.L. and Van Dop, C. (1989b). *J. Cardiovasc. Pharmacol.* **13**, 774–780.

Feldman, A.M., Tena, R.G., Kessler, P.D., Weisman, H.F., Schulman, S.P., Blumenthal, R.S., Jackson, D.G. and Van Dop, C. (1990). *Circulation* **81**, 1341–1352.

Finkel, M.S., Marks, E.S., Patterson, R.E., Speir, E.H., Steadman, K. and Keiser, H.R. (1986). *Am. J. Cardiol.* **57**, 1205–1206.

Fouad, F.M., Shimamatsu, K., Said, S.I. and Tarazi, R.C. (1986). *J. Cardiovasc. Pharmacol.* **8**, 398–405.

Fowler, M.B., Laser, J.A., Hopkins, G.L., Minobe, W. and Bristow, M.R. (1986). *Circulation* **74**, 1290–1302.

Freissmuth, M., Casey, P.J. and Gilman, A.G. (1989). *FASEB J.* **3**, 2125–2131.

Fujimoto, S., Dohi, Y., Aoki, K., Asano, M. and Matsuda, T. (1987). *Eur. J. Pharmacol.* **136**, 179–187.

Gertz, E.W. (1972). *Prog. Exp. Tumor Res.* **16**, 242–260.

Gilson, N., Bouanani, N.E.H., Corsin, A. and Crozatier, B. (1990). *Am. J. Physiol.* **258** (Heart Circ. Physiol. **27**), H634–H641.

Glaubiger, G. and Lefkowitz, R.J. (1977). *Biochem. Biophys. Res. Commun.* **78**, 720–725

Goldberg, L.I. (1972). *Pharmacol. Rev.* **24**, 1–29.

Grammas, P., Dereski, M.O., Diglio, C., Giacomelli, F. and Wiener, J. (1989). *J. Mol. Cell. Cardiol.* **21**, 807–815.

Hall, J.A., Kaumann, A.J. and Brown, M.J. (1990). *Circ. Res.* **66**, 1610–1623.

Hammond, H.K., White, F.C., Buxton, I.L.O., Saltzstein, P., Brunton, L.L. and Longhurst, J.C. (1987). *Am. J. Physiol.* **252** (Heart Circ. Physiol. **21**), H283–H290.

Harding, S.E., Jones, S.M., O'Gara, P., Vescovo, G. and Poole-Wilson, P.A. (1990). *Am. J. Physiol.* **259** (Heart Circ. Physiol. **28**), H1009–H1014.

Hausdorff, W.P., Caron, M.G. and Lefkowitz, R.J. (1990). *FASEB J.* **4**, 2881–2889.

Hazeki, O., Katada, T., Kurose, H. and Ui, M. (1983). In: *Physiology and Pharmacology of Adenosine Derivatives* (Daly, J.W., Kuroda, Y., Phillips, J.W., Shimizu, H. and Ui, M. eds), pp. 41–49. Raven Press, New York.

Heathers, G.P., Evers, A.S. and Corr, P.B. (1989). *J. Clin. Invest.* **83**, 1409–1413.

Hohl, C.M., Wetzel, S., Fertel, R.H., Wimsatt, D.K., Brierley, G.P. and Altschuld, R.A. (1989). *Am. J. Physiol.* **257** (Cell Physiol. **26**), C948–C956.

Homburger, F. (1979). *Ann. N. Y. Acad. Sci.* **317**, 2–17.

Hong, S.L. and Deykin, D. (1982). *J. Biol. Chem.* **257**, 7151–7154.

Horn, E.M. and Bilezikian, J.P. (1990). *Circulation* **82** (suppl. I), I-26–I-34.

Horn, E.M., Corwin, S.J., Steinberg, S.F., Chow, Y.K., Neuberg, G.W., Cannon, P.J., Powers, E.R. and Bilezikian, J.P. (1988). *Circulation* **78**, 1373–1379.

Jasmin, G. and Proschek, L. (1983). *Muscle Nerve* **6**, 408–415.
Jasmin, G. and Proschek, L. (1984). *Can. J. Physiol. Pharmacol.* **62**, 891–898.
Jasmin, G. and Solymoss B. (1975). *Proc. Soc. Exp. Biol. Med.* **149**, 193–198.
Jeffries, W.B., Yang, E. and Pettinger, W.A. (1988). *Hypertension* **12**, 80–88.
Jones, C.R., Molenaar, P. and Summers, R.J. (1989). *J. Mol. Cell. Cardiol.* **21**, 519–535.
Kagiya, T., Hori, M., Kitabatake, A., Inoue, M., Uchida, S. and Yoshida, H. (1985). *Jap. Circ. J.* **49**, 760.
Kammerling, J.J., Green, F.J., Watanabe, A.M., Inoue, H., Barber, M.J., Henry, D.P. and Zipes, D.P. (1987). *Circulation* **76**, 383–393.
Karliner, J.S. (1989). *J. Cardiovasc. Pharmacol.* **14** (suppl. 5), S-6–S-12.
Karliner, J.S. and Scheinman, M. (1988). *Am. J. Cardiol.* **62**, 1129–1130.
Karliner, J.S. and Simpson, P.C. (1988). *Basic Res. Cardiol.* **83**, 655–663.
Karliner, J.S., Barnes, P., Brown, M. and Dollery, C. (1980). *Eur. J. Pharmacol.* **67**, 115–118.
Karliner, J.S., Stevens, M., Woloszyn, W., Honbo, N. and Hoffman, J.I.E. (1986a). *J. Am. Coll. Cardiol.* **7**, 166A.
Karliner, J.S., Stevens, M., Grattan, M., Woloszyn, W., Honbo, N. and Hoffman, J.I.E. (1986b). *J. Am. Coll. Cardiol.* **8**, 349–356.
Karliner, J.S., Stevens, M.B., Honbo, N. and Hoffman, J.I.E. (1989). *J. Clin. Invest.* **83**, 474–481.
Kassis, S., Olasmaa, M., Terenius, L. and Fishman, P.H. (1987). *J. Biol. Chem.* **262**, 3429–3431.
Katoh, Y., Komuro, I., Takaku, F., Yamaguchi, H. and Yazaki, Y. (1990). *Circ. Res.* **67**, 235–239.
Kawabe, K., Watanabe, T.X., Shiono, K. and Sokabe, H. (1979). *Jap. Heart J.* **20**, 886–894.
Kennedy, R.H. and Donnelly, T.E. Jr (1982). *Gen. Pharmacol.* **13**, 231–239.
Kessler, P.D., Cates, A.E., Van Dop, C. and Feldman, A.M. (1989). *J. Clin. Invest.* **84**, 244–252.
Klein, I. and Levey, G.S. (1971). *J. Clin. Invest.* **50**, 1012–1015.
Kobayashi, A., Yamashita, T., Kaneko, M., Nishiyama, T., Hayashi, H. and Yamazaki, N. (1987). *J. Am. Coll. Cardiol.* **10**, 1128–1134.
Krawietz, W., Werdan, K. and Erdmann, E. (1982). *Biochem. Pharmacol.* **31**, 2463–2469.
Kurachi, Y., Nakajima, T. and Sugimoto, T. (1986). *Pflugers Arch.* **407**, 264–274.
Lambert, T.L., Kent, R.S. and Whorton, A.R. (1986). *J. Biol. Chem.* **261**, 15288–15293.
Lee, R.T., Brock, T.A., Tolman, C., Bloch, K.D., Seidman, J.G. and Neer, E.J. (1989). *FEBS Lett.* **249**, 132–142.
Lefkowitz, R.J., Hausdorff, W.P. and Caron, M.G. (1990). *Trends Pharmacol. Sci.* **11**, 190–194.
Levey, G.S. and Epstein, S.E. (1969). *Circ. Res.* **14**, 151–156.
Levine, M.A., Feldman, A.M., Robishaw, J.D., Ladenson, P.W., Ahn, T.G., Moroney, J.F. and Smallwood, P.M. (1990). *J. Biol. Chem.* **265**, 3553–3560.
Liang, C., Fan, T.-H.M., Sullebarger, J.T. and Sakamoto, S. (1989). *J. Clin. Invest.* **84**, 1267–1275.

Limas, C.J. (1979). *Biochim. Biophys. Acta* **588**, 174–178.

Limas, C.J. and Limas, C. (1984). *Circ. Res.* **55**, 524–531.

Limas, C.J. and Limas, C. (1987). *Am. J. Physiol.* **253** (Heart Circ. Physiol. 22), H904–H908.

Limas, C.J., Goldenberg, I.F. and Limas, C. (1989). *Circ. Res.* **64**, 97–103.

Limas, C.J., Limas, C., Kubo, S.H. and Olivari, M.T. (1990a). *Am. J. Cardiol.* **65**, 483–487.

Limas, C.J., Goldenberg, I.F. and Limas, C. (1990b). *Am. Heart J.* **119**, 1322–1328.

Lindenfeld, J., Crawford, M.H., O'Rourke, R.A., Levine, S.P., Montiel, M.M. and Horwitz, L.D. (1980). *Circulation* **62**, 704–711.

Longabaugh, J.P., Vatner, D.E., Vatner, S.F. and Homcy, C.J. (1988). *J. Clin. Invest.* **81**, 420–424.

Luetje, C.W., Tietje, K.M., Cristian, J.L. and Nathanson, N.M. (1988). *J. Biol. Chem.* **263**, 13357–13365.

Maisel, A.S., Motulsky, H.J. and Insel, P.A. (1985). *Science* **230**, 183–185.

Maisel, A.S., Motulsky, H.J. and Insel, P.A. (1986). *Circ. Res.* **60**, 108–112.

Maisel, A.S., Motulsky, H.J., Ziegler, M.G. and Insel, P.A. (1987). *Am. J. Physiol.* **253** (Heart Circ. Physiol. 22), H1159–H1166.

Maisel, A.S., Ziegler, M.G., Carter, S., Insel, P.A. and Motulsky, H.J. (1988). *J. Clin. Invest.* **82**, 2038–2044.

Maisel, A.S., Michel, M.C., Insel, P.A., Ennis, C., Ziegler, M.G. and Phillips, C. (1990a). *Circulation* **81**, 1198–1204.

Maisel, A.S., Ransnas, L.A. and Insel, P.A. (1990b). In: *Adrenergic Mechanisms in Myocardial Ischemia* (Heusch, G. and Ross, J. Jr, eds), Supplement to *Basic Res. Cardiol.* **85** (suppl. 1), pp. 47–56. Springer-Verlag, New York.

Makita, N. and Yasuda, H. (1990). *Basic Res. Cardiol.* **85**, 435–443.

Malbon, C.C., Rapiejko, P.J. and Mangano, T.J. (1985). *J. Biol. Chem.* **260**, 2558–2564.

Malbon, C.C., Rapiejko, P.J. and Watkins, D.C. (1988). *Trends Pharmacol. Sci.* **9**, 33–36.

Mancini, D.M., Frey, M.J., Fischberg, D., Molinoff, P.B. and Wilson, J.R. (1989). *Am. J. Cardiol.* **63**, 307–312.

Manning, A.S., Yellon, D.M., Coltart, D.J. and Hearse, D.J. (1981). *J. Mol. Cell. Cardiol.* **13**, 999–1009.

Marsh, J.D. and Sweeney, K.A. (1989). *Am. J. Physiol.* **256** (Heart Circ. Physiol. 25), H275–H281.

Marsh, J.D., Barry, W.H. and Smith, T.W. (1982). *J. Pharmacol. Exp. Ther.* **223**, 60–67.

Mattera, R., Pitts, B.J., Entman, M.L. and Birnbaumer, L. (1985). *J. Biol. Chem.* **260**, 7410–7421.

Michel, M.C., Kanczik, R., Khamssi, M., Knorr, A., Siegl, H., Beckeringh, J.J. and Brodde, O.E. (1989). *J. Cardiovasc. Pharmacol.* **13**, 421–431.

Michel, M.C., Brodde, O.E. and Insel, P.A. (1990). *Hypertension* **16**, 107–120.

Millanvoye, E., Freyss-Beguin, M., Baudouin-Legros, M., Paquet, J.-L., Marche, P., Durant, S. and Meyer, P. (1988). *J. Hypertension* **6** (suppl. 4), S369–S371.

Mügge, A., Reupcke, C. and Scholz, H. (1985). *Eur. J. Pharmacol.* **112**, 249–252.

Mukherjee, A., Wong, T.M., Buja, L.M., Lefkowitz, R.J. and Willerson, J.T. (1979). *J. Clin. Invest.* **64**, 1423–1428.

Mukherjee, A., Bush, L.R., McCoy, K.E., Duke, R.J., Hagler, H., Buja, L.M. and Willerson, J.T. (1982). *Circ. Res.* **50**, 735–741.

Muntz, K.H., Olson, E.G., Lariviere, G.R., D'Souza, S., Mukherjee, A., Willerson, J.T. and Buja, L.M. (1984). *J. Clin. Invest.* **73**, 349–357.

Murad, F. and Vaughan, M. (1969). *Biochem. Pharmacol.* **18**, 1053–1059.

Murakami, T. and Yasuda, H. (1986). *Biochim. Biophys. Res. Commun.* **138**, 1355–1361.

Murakami, T., Katada, T. and Yasuda, H. (1987). *J. Mol. Cell. Cardiol.* **19**, 199–208.

Myers, J.H. and Horwitz, L.D. (1978). *Circulation* **58**, 196–201.

Myers, M.G., Freeman, M.R., Juma, Z.A. and Wisenberg, G. (1979). *Am. Heart J.* **97**, 298–302.

Näbauer, M., Böhm, M., Brown, L., Diet, F., Eichhorn, M., Kemkes, B., Pieske, B. and Erdmann, E. (1988). *Eur. J. Clin. Invest.* **18**, 600–606.

Nanoff, C., Freissmuth, M., Tuisl, E. and Schütz, W. (1989). *J. Cardiovasc. Pharmacol.* **13**, 198–203.

Narayanan, T.K., Confer, R.A., Dennison, R.L. Jr, Anthony, B.L. and Aronstam, R.S. (1988). *Biochem. Pharmacol.* **37**, 1219–1223.

Neumann, J., Schmitz, W., Scholz, H., von Meyerinck, L., Döring, V. and Kalmar, P. (1988). *Lancet* **2** (8617), 936–937.

Nichols, A.J., Motley, E.D. and Ruffolo, R.R. Jr (1988). *Eur. J. Pharmacol.* **145**, 345–349.

Noren, G.R., Staley, E.F., Jankus, E.F. and Stevenson, J.F. (1971). *Arch. Abt. A. Pathol. Anat.* **352**, 285–295.

Olson, H.G., Miller, R.R., Amsterdam, E.A., Wood, M., Brocchini, R. and Manson, D.T. (1975). *Am. J. Cardiol.* **35**, 162.

Ransnas, L.A. and Insel, P.A. (1988). *J. Biol. Chem.* **263**, 9482–9485.

Ransnas, L.A. and Insel, P.A. (1989). *Anal. Biochem.* **176**, 185–190.

Reithmann, C. and Werdan, K. (1988). *Eur. J. Pharmacol.* **154**, 99–104.

Reithmann, C. and Werdan, K. (1989). *Naunyn-Schmiedeberg's Arch. Pharmacol.* **339**, 138–144.

Reithmann, C., Gierschik, P., Sidiropoulos, D., Werdan, K. and Jakobs, K.H. (1989). *Eur. J. Pharmacol.* **172**, 211–221.

Robinson, T.W. and Giri, S.N. (1986). *Biochim. Biophys. Res. Commun.* **136**, 745–752.

Roncari, D.A. and Murthy, V.K. (1975). *J. Biol. Chem.* **250**, 4134–4138.

Roth, D.A., Urasawa, K., Leiber, D., Insel, P.A. and Hammond, H.K. (1992). *FEBS Lett.* (in press).

Saragoça, M. and Tarazi, R.C. (1981a). *Hypertension* **3**, 380–385.

Saragoça, M.A. and Tarazi, R.C. (1981b). *Hypertension* **3** (suppl. II), II-171–II-176.

Sato, K. (1990). *Hokkaido J. Med. Sci.* **65**, 466–473.

Schnabel, P., Böhm, M., Gierschik, P., Jakobs, K.H. and Erdmann, E. (1990). *J. Mol. Cell. Cardiol.* **22**, 73–82.

Schömig, A., Dart, A.M., Dietz, R., Mayer, E. and Kübler, W. (1984). *Circ. Res.* **55**, 689–701.

Schömig, A., Fischer, S., Kurz, T., Richardt, G. and Schömig, E. (1987). *Circ. Res.* **60**, 194–205.

Schubert, B., VanDongen, A.M.J., Kirsch, G.E. and Brown, A.M. (1989). *Science* **245**, 516–519.

Schultz, K.D., Fritschka, E., Pauliks, L.B., Philipp, T. and Distler, A. (1989). *J. Hypertension* **7** (suppl.), S142–S143.

Schwinger, R.H., Böhm, M. and Erdmann, E. (1990). *J. Cardiovasc. Pharmacol.* **15**, 692–697.

Seamon, K.B. and Daly, J.W. (1986). *Adv. Cyclic Nucleotide Protein Phosphorylation Res.* **20**, 1–150.

Sen, L., Liang, B.T., Colucci, W.S. and Smith, T.W. (1990). *Circ. Res.* **67**, 1182–1192.

Simon, M.I., Strathmann, M.P. and Gautam, N. (1991). *Science* **252**, 802–808.

Slome, R. (1973). *Lancet* **1**, 156.

Smith, H.J. and Nuttall, A. (1985). *Cardiovasc. Res.* **19**, 181–186.

Smrcka, A.V., Hepler, J.R., Brown, K.O. and Sternweis, P.C. (1991). *Science* **251** (4995), 804–807.

Sobierai, J., Eastman, E., Chobanian, A.V. and Brecher, P. (1989). *Clin. Res.* **37**, 604A.

Staley, N.A., Eizig, S., Noren, G.R., Surdy, J.E. and Elsperger, J. (1987). *Am. J. Physiol.* **252** (Heart Circ. Physiol. 21), H334–H339.

Sterin-Borda, L., Cantore, M., Pascual, J., Borda, E., Cassio, P., Arana, R. and Passeron, S. (1986). *Int. J. Immunopharmacol.* **8**, 581–588.

Sterin-Borda, L., Leiros, C.P., Wald, M., Cremaschi, G. and Borda, E. (1988). *Clin. Exp. Immunol.* **74**, 349–354.

Stiles, G.L. and Lefkowitz, R.J. (1981). *Life Sci.* **28**, 2529–2536.

Stiles, G.L., Caron, M.G. and Lefkowitz, R.J. (1984). *Pharmacol. Rev.* **64**, 661–743.

Strasser, R.H., Krimmer, J., Braun-Dullaeus, R., Marquetant, R. and Kübler, W. (1990). *J. Mol. Cell. Cardiol.* **22**, 1405–1423.

Sugiyama, T., Yoshizumi, M., Takaku, F., Urabe, H., Tsukakoshi, M., Kasuya, T. and Yazaki, Y. (1986). *Biochem. Biophys. Res. Commun.* **141**, 340–345.

Susanni, E.E., Manders, W.T., Knight, D.R., Vatner, D.E., Vatner, S.F. and Homcy, C.J. (1989). *Circ. Res.* **65**, 1145–1150.

Sutherland, E.W., Rall, T.W. and Menon, T. (1962). *J. Biol. Chem.* **237**, 1220–1243.

Tenner, T.E. Jr (1983a). *Eur. J. Pharmacol.* **92**, 91–97.

Tenner, T.E. Jr (1983b). *Life Sci.* **33**, 2291–2299.

Tse, J., Powell, J.R., Baste, C.A., Priest, R.E. and Kuo, J.F. (1979). *Endocrinology* **105**, 246–255.

Tse, J., Wrenn, R.W. and Kuo, J.F. (1980). *Endocrinology* **107**, 6–16.

Uehara, Y., Ishii, M., Ishimitsu, T. and Sugimoto, T. (1988). *Hypertension* **11**, 28–33.

Urasawa, K., Sato, K., Murakami, T., Kawaguchi, H. and Yasuda, H. (1990a). *Life Sci.* **47**, 1761–1767.

Urasawa, K., Igarashi, Y., Kawaguchi, H. and Yasuda, H. (1990b). *Jap. Circ. J.* **54**, 980.

Vatner, D.E., Vatner, S.F., Fujii, A.M. and Homcy, C.J. (1985). *J. Clin. Invest.* **76**, 2259–2264.

Vatner, D.E., Lee, D.L., Schwarz, K.R., Longabaugh, J.P., Fujii, A.M., Vatner, S.F. and Homcy, C.J. (1988a). *J. Clin. Invest.* **81**, 1836–1842.

Vatner, D.E., Knight, D.R., Shen, Y.T., Thomas, J.X. Jr, Homcy, C.J. and Vatner, S.F. (1988b). *J. Mol. Cell. Cardiol.* **20**, 75–82.

Vatner, D.E., Vatner, S.F., Nejima, J., Uemura, N., Susanni, E.E., Hintze, T.H. and Homcy, C.J. (1989). *J. Clin. Invest.* **84**, 1741–1748.

Vatner, D.E., Young, M.A., Knight, D.R. and Vatner, S.F. (1990). *Am. J. Physiol.* **258** (Heart Circ. Physiol. 27), H140–H144.

Vinicor, F., Higdon, G. and Clark, C.M. Jr (1977). *Endocrinology* **101**, 1071–1077.

Wagner, J.A., Reynolds, I.J., Weisman, H.F., Dudeck, P., Weisfeldt, M.L. and Snyder, S.H. (1986). *Science* **232**, 515–517.

Webb, J.G., Newman, W.H., Walle, T. and Daniell, H.B. (1981). *J. Cardiovasc. Pharmacol.* **3**, 622–635.

Whitmer, J.T., Kumar, P. and Solaro, R.J. (1988). *Circ. Res.* **62**, 81–85.

Wildenthal, K. (1972). *J. Clin. Invest.* **51**, 2702–2709.

Will-Shahab, L., Krause, E.G., Bartel, S., Schulze, W. and Küttner, I. (1985). *J. Cardiovasc. Pharmacol.* **7** (suppl 5), S23–S27.

Will-Shahab, L., Küttner, I. and Warbanow, W. (1986a). *Biomed. Biochim. Acta* **45**, S199–S204.

Will-Shahab, L., Krause, E.G., Schulze, W. and Bartel, S. (1986b). *Cor Vasa* **28**, 107–113.

Wolff, A.A., Hines, D.K. and Karliner, J.S. (1989). *Am. J. Physiol.* **257** (Heart Circ. Physiol. 26), H1032–H1036.

Wong, S.K.F. and Martin, B.R. (1983). *Biochem. J.* **216**, 753–759.

Yatani, A., Codina, J., Brown, A.M. and Birnbaumer, L. (1987a). *Science* **235**, 207–211.

Yatani, A., Codina, J., Imoto, Y., Reeves, J.P., Birnbaumer, L. and Brown, A.M. (1987b). *Science* **238**, 1288–1292.

Zidek, W., Vetter, H., Dorst, K.G., Zumkley, H. and Losse, H. (1982). *Clin. Sci.* **63** (suppl.), 41–43.

CHAPTER FOUR

Alterations in G-protein-mediated Cell Signalling in Diabetes Mellitus

CHRISTOPHER J. LYNCH* and JOHN H. EXTON[†]
* *Department of Cellular and Molecular Physiology,
The Milton S. Hershey Medical Center, The Pennsylvania State University College of Medicine, Hershey,
PA 17033, USA*
[†] *Department of Molecular Physiology and Biophysics, Vanderbilt University Medical School,
Nashville, TN 37232, USA*

1 INTRODUCTION

1.1 Diabetes mellitus in humans

Diabetes mellitus is classified into two major types. The more severe form of diabetes (type I or insulin-dependent) usually appears before adulthood. It is associated with ketosis in the untreated state and accounts for 10–20% of diabetics in the developed countries. It is associated with increased frequency of certain HLA antigens. Insulin is virtually absent from the plasma and glucagon is elevated.

The other form of diabetes (type II or non-insulin dependent) is milder and occurs predominantly in adults. It is more heterogeneous, but is usually associated with tissue insensitivity to insulin. Most subjects are obese and show insulin resistance, but a small fraction (15%) are non-obese. All exhibit an abnormal insulin response to glucose. In rare instances, the diabetic state may be due to the production of abnormal forms of insulin.

Untreated type I diabetics show hyperglycaemia and hyperketonaemia, with consequent glycosuria and polyuria due to osmotic diuresis. There is weight loss despite increased appetite. If the insulin deficiency is severe and acute, there is ketoacidosis, hyperosmolality and dehydration. If uncompensated, these lead to nausea and vomiting, hyperventilation, loss of consciousness and circulatory collapse. The

G-Proteins
ISBN 0-12-497515-1

hyperglycaemia is the result of overproduction of glucose by the liver, due to enhanced gluconeogenesis, and to under-utilization of glucose by peripheral tissues. The hyperketonaemia is due to enhanced release of free fatty acids from adipose tissue which are taken up by the liver and converted to ketone bodies. The stimulations of gluconeogenesis, lipolysis and ketogenesis are due directly and indirectly to the deficiency of insulin and the excess of glucagon. This is because glucagon is a stimulant to gluconeogenesis, lipolysis and ketogenesis in diabetic subjects, whereas insulin normally acts to inhibit these processes and to reduce the supply of gluconeogenic substrates to the liver.

Type II diabetics show milder hyperglycaemia than type I diabetics and no ketoacidosis. They may be relatively asymptomatic initially. However, when the disease is established, they, like type I diabetics, show polyuria, thirst, blurred vision, paraesthesiae, fatigue, susceptibility to infections and problems with gestation and delivery. Elevation of glycosylated haemoglobin (haemoglobin A_{1c}) is observed in both forms of diabetes, and thickening of the basement membranes of capillaries and atherosclerosis occur as chronic complications leading to diabetic retinopathy, renal glomerulosclerosis, peripheral and autonomic neuropathy, and increased incidence of myocardial infarction, stroke and peripheral gangrene.

Type II diabetics have an impairment of insulin secretion since their insulin levels are lower than those observed in normal individuals with comparable hyperglycaemia. Furthermore, they usually do not display the acute (first phase) insulin response to intravenous glucose. However, the major defect is peripheral insulin resistance. The molecular basis of this remains unclear, but there is evidence for a reduction in cellular insulin receptors and also a post-receptor defect in insulin action. Part of the latter may be due to a deficiency of glucose transporters.

Hepatic glucose output is increased in type II diabetics and this is probably due to the elevation of plasma glucagon that is consistently observed. The increase in glucagon suggests that the A-cells of the pancreas do not respond to the normal suppressive effects of glucose and insulin. An additional factor in the increased glucose production may be an increased flow of gluconeogenic substrates to the liver.

1.2 Animal models of diabetes

One of the inital experimental models of diabetes was that induced by subtotal pancreatectomy. This is a difficult operation to perform and the animals generally have problems in surviving. Two other forms of experimental diabetes are those induced by alloxan and streptozotocin, which destroy the B-cells of the pancreas, but are toxic to other cells. An acute form of insulin deficiency can also be produced by injection of anti-insulin antiserum.

Genetically determined forms of diabetes provide models of the disease that are better than those described above. The db/db mouse (Herberg and Coleman, 1977), which is a variant of the C57BL/K$_s$J strain, shows progressive obesity which is transmitted as a recessive trait. The hyperadiposity is initially associated with hyperinsulinism and then with insulin resistance which becomes pronounced. The db/db mouse may thus be thought of as a model for human type II diabetes. Other insulin-resistant rodents with transient or stable non-ketotic diabetes include the ob/ob mouse, the yellow and KK 'diabese' mice, the PBB/Ld mouse, and the diabetic obese Zucker rat (review: Shafrir, 1990).

Another useful model is the BB (Bio-breeding) rat (Nakhooda *et al.*, 1977) and the non-obese diabetic (NOD) mouse (Fujita *et al.*, 1982) which show spontaneous diabetes without obesity. In the BB rat, the disease manifests itself between two and four months of age in about 30% of the animals.

In both models there is acute pancreatic insulitis and insulin deficiency. Unless the animals are injected with insulin, they die in ketoacidosis, and thus resemble type I human diabetics.

2 EVIDENCE OF DISORDERS OF G-PROTEIN SIGNALLING SYSTEMS IN DIABETES MELLITUS

As described above, human diabetics show insulin deficiency and/or insulin resistance. Plasma glucagon is either normal or elevated (Unger and Orci, 1981). Elevations in plasma glucagon have also been reported in animal models of diabetes mellitus (Laube *et al.*, 1973; Kampa *et al.*, 1981; Chang *et al.*, 1977). Plasma noradrenaline and adrenaline concentrations are normal in controlled diabetics but are considerably elevated during ketoacidosis (Christensen, 1974; Cryer, 1980). In poorly controlled diabetics and in alloxan-diabetic dogs the catecholamine response to exercise is abnormally exaggerated (Christensen, 1974; Tamborlane *et al.*, 1979; Wasserman *et al.*, 1985). In the long term, autonomic neuropathy may occur, especially with poor insulin control, though the parasympathetic supply seems to be more sensitive than the sympathetic (review: Fein and Scheuer, 1990). Lastly, plasma growth hormone concentrations have been reported to be increased in juvenile-onset diabetics (Hayford *et al.*, 1980) and the growth hormone response to exercise is also above normal in poorly controlled patients (Hansen, 1971; Tamborlane *et al.*, 1979; Topper *et al.*, 1985). Because glucagon, catecholamines and insulin are involved in the regulation of cAMP levels in liver and adipose tissue as well as other tissues, changes in the cellular level of this nucleotide are observed in diabetes. Thus cAMP has been reported to increase in the livers of alloxan- and streptozotocin-diabetic rats and following the injection of anti-insulin serum (Jefferson *et al.*, 1968; Pilkis *et al.*, 1974; Goldberg *et al.*, 1969). As expected, these changes are reversed by insulin treatment. BB/W (Bio/Breeding Worcester) rats also have a two-fold elevation in hepatic cAMP (Appel *et al.*, 1981). On the other hand, cAMP levels in the livers of young db/db mice are not elevated (Chan *et al.*, 1975; Levilliers *et al.*, 1978) as found for other types of obese animals (Exton, 1980).

There have been few measurements of cAMP in the adipose tissue of animals with experimental diabetes. In epididymal adipose tissue of rats treated with alloxan or anti-insulin antiserum, cAMP has been reported to be elevated several-fold (Schimmel, 1976). On the other hand, a decrease has been reported in db/db mice (Levilliers *et al.*, 1978), but this is largely due to the increase in fat content.

The marked increase in cAMP in the livers of animal models of type I diabetes probably contributes to the increase in gluconeogenesis in such animals because of the induction of key gluconeogenic enzymes such as phosphoenolpyruvate carboxykinase and changes in the phosphorylation state of pyruvate kinase and 6-phosphofructo-2-kinase (Granner and Pilkis, 1990). It may also play a role in the altered glycogen metabolism and enhanced ketogenesis because of increased phosphorylation and altered activities of glycogen phosphorylase, glycogen synthase and acetyl CoA carboxylase. The elevation in cAMP in the adipocytes of these animals is probably a factor in the enhanced mobilization of free fatty acids and resultant ketogenesis.

The plasma concentrations of cAMP-stimulating hormones and their exaggerated secretion in response to exercise can be normalized with careful insulin control (Tamborlane *et al.*, 1979). The expectation, *a priori*, might be that long-term exposure to elevated levels of anti-insulin hormones would result in decreased tissue responsiveness to these hormones by activating well-known desensitization mechanisms. Alternatively, in well-controlled diabetics one might expect no changes in tissue responsiveness. However, this is contrary to the observation, well known to clinical diabetologists, that stress responses are greater in diabetics than in non-diabetics, *despite ongoing insulin treatment* (Bondy and Felig, 1974; Shamoon *et al.*, 1980). Thus salbutamol administration produces a significantly larger lipolytic and gluconeogenic response in juvenile diabetics as compared to non-diabetic controls (Wager *et al.*, 1981) and the glucose response of diabetics to exogenously administered adrenaline and glucagon is exaggerated even in insulin-infused juvenile diabetics (Shamoon *et al.*, 1980). This suggests an increased rather than decreased response of the liver and fat to the anti-insulin hormones. Presumably this altered response is the result of changes in G-protein-mediated cell signalling in these tissues.

Another tissue in which cAMP metabolism has been examined in diabetic animals is the heart. Alloxan-diabetic rats have unaltered basal levels of cAMP in their hearts, but the increase with β-adrenergic agonists is diminished (Ingebretsen *et al.*, 1981; Miller *et al.*, 1981; Miller, 1984). Despite this, the activation of glycogen phosphorylase due to β-adrenergic stimulation is increased. In the hearts of BB/Wor rats, basal and adrenaline-stimulated cAMP levels are not increased and the activation of cAMP-dependent protein kinase is unaltered (Miller, 1983). However, as seen in alloxan-diabetic rats, phosphorylase activation by the catecholamine is enhanced.

The data reported above indicate that there are changes in cAMP and cAMP-mediated responses in liver, adipose tissue and heart of animals and humans with diabetes mellitus. The possibility that this involves alterations in the G-protein-mediated cell signalling is discussed below.

3 CHANGES IN G-PROTEIN-MEDIATED SIGNALLING IN DIABETES MELLITUS

3.1 Liver

The action of catecholamines on the liver is mediated by α_1-adrenergic or β_2-adrenergic receptors (Exton *et al.*, 1978; Exton, 1987). The α_1-adrenergic receptors couple to two pertussis toxin-insensitive G-proteins of the G_q class (Taylor et al., 1991; Wange et al., 1991) involved in phosphoinositide hydrolysis, while β_2-adrenergic receptors couple to G_s, which in turn is involved at least in adenylyl cyclase activation. The predominance of α_1- or β_2-adrenergic responsiveness is species dependent and at least in the rat is also dependent on age and sex. Glucagon's actions are probably only mediated by G_s in the liver. Thus, while it is true that high concentrations of glucagon are capable of stimulating inositol 1,4,5-trisphosphate (IP_3) formation and calcium mobilization, this action is mimicked by forskolin as well as cAMP analogues and may therefore be secondary to the increase in cAMP. It has been proposed that it may be due to phosphorylation and activity changes in the G-proteins coupled to phosphatidylinositol 4,5-bisphosphate (PIP_2) hydrolysis or their associated receptors (Blackmore and Exton, 1986). Additionally, cAMP-dependent phosphorylation may sensitize the intracellular Ca^{2+} stores to the action of IP_3 (Burgess et al., 1991).

Hepatocytes from rats made acutely or chronically diabetic with streptozotocin show increased levels of cytosolic Ca^{2+} (Studer and Ganas, 1989). The Ca^{2+} and phosphorylase responses of these cells to adrenaline are impaired, although there is no evidence of altered α_1-adrenergic receptors. The Ca^{2+} responses to vasopressin, angiotensin II and glucagon are also reduced. Since the IP_3 responses to adrenaline and vasopressin are depressed, whereas the ability of IP_3 to release intracellular Ca^{2+} is not impaired, as assessed in saponin-permeabilized cells (Studer and Ganas, 1989), it appears that one of the defects is either reduced IP_3 formation or enhanced IP_3 degradation. In the case of the former, this could be due to an impairment in the G-proteins that couple the receptors to PIP_2 phospholipase C. However, no data are available on the levels or activities of the phospholipase or the G_q class of G-proteins in the tissues of diabetic animals.

In diabetes mellitus the liver turns from a glucose-storing organ to a glucose producing organ. This is a consequence of decreased plasma insulin, increased plasma levels of the anti-insulin stress hormones, noradrenaline, adrenaline, glucagon and cortisol (Tamborlane et al., 1979; Christensen, 1974; Cryer, 1980; Unger and Orci, 1981) and increased hepatic responsiveness to the anti-insulin hormones even in treated diabetes and insulin-infused diabetic animals (Bondy and Felig, 1974; Shamoon et al., 1980). Thus a uniformly observed event in animal models of juvenile diabetes mellitus is increased in vivo hepatic levels of basal and/or hormone stimulated cAMP levels (Jefferson et al., 1968; Pilkis et al., 1974; Wagle et al., 1975; Shikama and Ui, 1976;

Seitz *et al.*, 1979; Appel *et al.*, 1981). At least part of this increased responsiveness to catecholamines and glucagon is due to decreased phosphodiesterase activity (Senft *et al.*, 1968; Pilkis *et al.*, 1974; Solomon *et al.*, 1986). However, it has long been suspected that changes in the more proximal steps in the cell-signalling pathway exist as well. Efforts to address this question with chemically diabetic rats have yielded discrepant results. For instance, in a recent study Gawler *et al.* (1987) reported that levels of α_i (α subunit of G_i) were abolished in hepatocytes from streptozotocin and alloxan diabetics.

These findings were confirmed by Bushfield *et al.* (1990a) who observed that $\alpha_i 2$ and $\alpha_i 3$ and also the 42-kDa form of α_s (α subunit of G_s), but not the 45-kDa form, were markedly decreased in hepatocyte membranes from streptozotocin-diabetic rats. Immunoblot analysis also indicated that the catalytic subunit of adenylyl cyclase was increased in membranes from hepatocytes and whole liver. Lynch *et al.* (1989) repeated the studies of Gawler *et al.* only to find that α_i was not changed either in whole hepatocytes or plasma membranes, but, rather, both species of α_s were increased. Either finding would contribute to the increased responsiveness to anti-insulin hormones in diabetes. But which observations are correct?

Bushfield *et al.* (1990c) proposed that the source of the plasma membranes (i.e. hepatocytes versus intact liver) used in the two studies accounted for the differences in the observations. However, this explanation does not account for the finding of equivalent levels of pertussis toxin-stimulated ADP-ribosylation of α_i in parenchymal hepatocytes from normal and streptozotocin-treated rats reported by Lynch *et al.* (1989). Furthermore, the explanation would require the plasma membranes studied by Lynch *et al.* to have been derived largely from non-parenchymal cells, which seems very unlikely since they were enriched in receptors and enzymes present in the plasma membranes of parenchymal cells.

Such discrepancies would seem surprising; however, an analysis of the extensive literature on the early steps in glucagon and catecholamine action in the chemically diabetic liver *in vitro* indicates incongruity as the only reproducible finding. For instance, studies of chemically diabetic rat livers have revealed either increased, decreased, or unchanged numbers of glucagon receptors (Srikant *et al.*, 1977; Bhathena *et al.*, 1978; Chamras *et al.*, 1980; Soman and Felig, 1978) as well as increased, decreased or equal changes in basal, GTP-, fluoride- or hormone-sensitive adenylyl cyclase activity in liver plasma membranes from chemically diabetic rats (Pilkis *et al.*, 1974; Srikant *et al.*, 1977; Soman and Felig, 1978; Chamras *et al.*, 1980; Israelsson and Tengrup, 1980; Allgayer *et al.*, 1982; Dighe *et al.*, 1984; Srivastava and Anand-Srivastava, 1985; Gawler *et al.*, 1987; Lynch *et al.*, 1989). We believe that a reasonable explanation for these discrepancies relates to the mechanism of action of the diabetogenic agents, especially streptozotocin.

The commonly held view that the diabetogenic drugs alloxan and streptozotocin exclusively target the pancreatic β-cells is probably not true. Streptozotocin contains both nitrosourea and deoxyglucose residues. The latter is involved in transporting the drug into pancreatic β-cells using the β-cell hexose transporter, presumably the GLUT 2 transporter. The nitrosourea moiety is involved in cell toxicity. The distribution of GLUT 2 transporters as well as other glucose transporters has been described (Bell *et al.*, 1990). If it is presumed that streptozotocin enters cells via the pancreatic β-cell GLUT 2 hexose transporter, a reasonable assumption would be that the drug could enter and damage other tissues containing this transporter, e.g. liver, kidney and intestine (Bell *et al.*, 1990). Indeed, preclinical evaluation of streptozotocin as an anti-tumour medication listed hepatotoxicity, nephrotoxicity, nausea, vomiting and diarrhoea as toxic side effects (Weiss, 1982).

Alloxan is thought to act by inhibiting glucokinase (hexokinase IV) which is present in the pancreatic β-cells and also in the liver. Because of this, it interferes with glucose-stimulated insulin release and glycolytic flux in the β-cell. Furthermore, glucose analogues block the toxic effects of alloxan *in vivo* and *in vitro* (reviews: Lenzen and Panten, 1988; Shafrir, 1990). These findings indicate that both drugs may have side effects that may depend on the fed or fasted state of the animal (not usually reported in the above-cited literature) as well as the dose, route of drug administration (intraperitoneal or intravenous) and stereoisomeric content of the drugs (not always reported by the vendor). In view of the above considerations, it is likely that the changes in G-proteins, hormone receptors and adenylyl cyclase might be different in a liver from a diabetic animal as compared to a liver from a diabetic animal recovering from variable levels of acute cytotoxic insult.

In the naturally occurring diabetes of the BB/Wor rat, basal and glucagon-, GTP- and cholera toxin-stimulated adenylyl cyclase activities were increased (Lynch *et al.*, 1989), while the ability of angiotensin to inhibit adenylyl cyclase (Jard *et al.*, 1981; Crane *et al.*, 1982) was not influenced. Cyclic AMP phosphodiesterase activity was decreased

(Solomon *et al.*, 1986). In the study of Lynch *et al.*, decreased pertussis toxin-catalysed [^{32}P]ADP-ribosylation of total α_i was observed but levels of α_i were unchanged when measured by immunoblotting. Cholera toxin-catalysed [^{32}P]ADP-ribosylation of both hepatic α_s species was increased and increased levels of α_s were also measured by immunoblotting. Similar results were recently reported in humans with type II diabetes mellitus by Caro *et al.* (1991). In addition, Caro *et al.* showed that the ability of insulin to inhibit pertussis toxin-mediated [^{32}P]ADP ribosylation of G_i (Rothenberg and Kahn, 1988) was attenuated in liver from humans with diabetes mellitus. However, the physiological relevance of this phenomenon remains to be determined (review: Exton, 1991). In conclusion, in the absence of any hormone receptor measurements during naturally occurring diabetes, it appears that the increased hepatic responsiveness to anti-insulin hormones is at least partly the result of increased G_s/G_i ratios and adenylyl cyclase, and decreased cAMP phosphodiesterase activity.

3.2 Adipose tissue

3.2.1 Insulin-dependent diabetes and animal models of type I diabetes

In animal models of type I diabetes and humans with this disease increased mobilization of fatty acids results from insulin deficiency and increased plasma levels of the anti-insulin hormones. These stimulate lipolysis by raising intracellular levels of cAMP (Burns *et al.*, 1972). Compared with non-diabetics, *in vivo* adipose tissue cAMP levels are elevated in the basal state and in response to stress in chemically diabetic rats (Schimmel, 1976; Kampa and Frascella, 1977). This is in agreement with findings in juvenile- and adult-onset diabetics requiring insulin (Engfeldt *et al.*, 1982). Both human and rat adipocytes contain β_1-, β_2- and β_3-adrenergic receptors which mediate the positive effects of catecholamines on lipolysis through α_s and adenylyl cyclase (reviews: Fain and Garcia-Sainz, 1983; Zaagsma and Nahorski, 1990). β_3-Adrenergic receptors are unique in their inability to be antagonized by propranolol (Zaagsma and Nahorski, 1990). Glucagon receptors in fat similarly couple to α_s and adenylyl cyclase. In contrast the mechanism by which growth hormone stimulates lipolysis remains to be elucidated (e.g. Goodman, 1970; Sengupta *et al.*, 1981; Solomon *et al.*, 1990), although it appears to require protein synthesis. Adipocytes also contain α_2-adrenergic receptors which inhibit adenylyl cyclase (Fain and Garcia-Sainz, 1983), presumably by activating α_i2 and increasing the levels of free β,γ subunits (Wong et al., 1991). Thus the magnitude of the lipolytic response of fat cells to catecholamines depends on the ratio of β- to α-adrenergic receptors. This ratio varies with species, fat cell size, adipose tissue location (e.g. abdominal or gluteal) and sex, as well as the presence of hormones (insulin, prostaglandins) and paracrine agents (adenosine, nicotinic acid) which have counter-regulatory effects on lipolysis (review: Arner, 1988).

Alterations in G-protein-mediated signal transduction pathways in diabetes were suggested by in vitro experiments in which adipocytes from diabetic rats or humans displayed increased sensitivity to hormones in terms of stimulated lipolysis and/or cellular levels of cAMP when compared with fat cells from non-diabetic controls (Zapf et al., 1975, 1978; Schimmel, 1976; Arner and Östman, 1976; Arner et al., 1979, 1981; Chiappe de Cingolani, 1983, 1986a; Solomon et al., 1985). Studies of adipocyte adenylyl cyclase have also revealed increased sensitivity to catecholamines in plasma membranes from chemically diabetic animals (Zumstein et al., 1980; Chiappe de Cingolani, 1986a). The changes in adenylyl cyclase are associated with increased numbers (or sensitivity) of adipose tissue β-adrenergic and growth hormone receptors in streptozotocin and human diabetes (Madar, 1984; Chiappe de Cingolani, 1986b; Wahrenberg et al., 1989, 1990; Solomon et al., 1990; Lacasa et al., 1983). In contrast glucagon receptor numbers are decreased (Mayor and Calle, 1988; Sato et al., 1989). However, the increased catecholamine and growth hormone sensitivity is not associated with any changes in immunoreactive α_s, α_i1 or α_i2, either in rats (Saggerson, 1989; Strassheim et al., 1990) or insulinopenic patients (Ohisalo et al., 1989). Interestingly, increases in adipocyte α_i1 and α_i3 mRNA levels and α_i3 immunoreactivity have been observed in streptozotocin-diabetic rats. However, the signal transduction pathway to which α_i3 couples has yet to be elucidated for the fat cell (α_i3 is capable of stimulating atrial K^+ channels (Codina et al., 1988; Mattera et al., 1989)). While changes in the amounts of immunoreactive α_s or $\alpha_i1,2$ may not explain the increased responsiveness of adenylyl cyclase to catecholamines in diabetes, Strassheim et al. (1990) have reported that diabetes abolished GTP-dependent (i.e. tonic) but not hormone-stimulated G_i function in streptozotocin-diabetic rat adipocyte plasma membranes. Changes in α_i2 phosphorylation state have been proposed to explain this change (Bushfield et al., 1990c).

In conclusion, the increased adipose tissue sensitivity to hormones which elevate cAMP and stimulate lipolysis observed in diabetes is probably the result of changes in receptor binding parameters, cAMP phosphodiesterase and possibly decreased tonic (but not hormonal) inhibition of adenylyl cyclase by a partially dysfunctional α_i2.

3.2.2 Animal models of non-insulin dependent diabetes

Type II diabetes in animal models and humans is normally associated with some degree of obesity. The insulin resistance which develops in peripheral tissues in these diabetics develops at a much slower rate in the adipose tissue and results in preferential shunting of substrate to fat, resulting in adipose tissue hypertrophy and hyperplasia (review: Caro *et al.*, 1990). The term diabesity has been proposed to describe this disorder (Shafrir, 1990) which is associated with impaired anti-insulin hormone sensitivity of adipose tissue. This is discussed in Chapter 6.

3.3 Heart

Minute to minute control of the heart is mediated primarily by the autonomic nervous system. Thus the exaggerated catecholamine response to stress in diabetes can be expected to have effects on adrenergic responsiveness in the heart. In contrast, in chronic poorly controlled diabetes, autonomic neuropathy is a problem. Autonomic neuropathy in the heart reveals itself clinically as variability in the beat to beat interval. The parasympathetic nervous system appears to be damaged earlier than the sympathetic nervous system in diabetes (review: Fein and Scheuer, 1990). β_1-Adrenergic receptors increase the heart rate in the SA node and increase conduction velocity, automaticity and/or contractility in the atria, AV node, His-Purkinje system and ventricles (Lefkowitz *et al.*, 1990). The β_1 adrenergic receptors couple to the two α_s subunit species found in the heart (α_s45 and α_s52), which in turn are at least capable of activating adenylyl cyclase and voltage sensitive Ca^{2+} channels (review: Robishaw and Foster, 1989). Muscarinic receptors in the heart inhibit adenylyl cyclase, presumably via α_i2 (Wong *et al.*, 1991), and stimulate atria K^+ channels via α_i3 (Codina *et al.*, 1988; Mattera *et al.*, 1989). α_1-Adrenergic receptors have a positive inotropic effect in the heart which may be mediated by a pertussis toxin-sensitive G-protein. In diabetic humans with some degree of autonomic neuropathy and in diabetic rats and rabbits, blunted cardiac responses to isoproterenol are observed (Zola *et al.*, 1988; Savarese and Berkowitz, 1979; Berlin *et al.*, 1986; Kenno and Severson, 1985; Atkins *et al.*, 1985; Ramanadham *et al.*, 1986; Ramanadham and Tenner, 1986; Almira and Misbin, 1989). This is associated with decreased β_1-adrenergic cAMP responses (Ingebretsen *et al.*, 1981; Gotzche, 1983) in chemically diabetic but not BB/Wor rats (Miller, 1983). β-Adrenergic stimulation of adenylyl cyclase is less in microsomes from streptozotocin- and alloxan-diabetic rats while guanine nucleotide, fluoride and forskolin responses appear to be unaltered (Menahan *et al.*, 1977; Savarese and Berkowitz, 1979; Chatelain *et al.*, 1983; Gotzche, 1983; Smith *et al.*, 1984; Michel *et al.*, 1985; Sundaresen *et al.*, 1984; Atkins *et al.*, 1985; Alishio *et al.*, 1988). Not surprisingly, then, changes in α_s or α_i immunoreactivity or mRNA levels are unaltered (Bushfield *et al.*, 1990c). These findings suggest that the alterations in β-adrenergic responsiveness are probably mediated at the receptor level. Indeed, β-adrenergic receptors are decreased in the hearts of chemically diabetic rats (Savarese and Berkowitz, 1979; Ingebretsen *et al.*, 1983; Ramanadham and Tenner, 1983; Atkins *et al.*, 1985; Sundaresan, 1986; Bitar *et al.*, 1987; Ramanadham and Tenner, 1987; Nishio *et al.*, 1988), but not in diabetic rabbits (Zola *et al.*, 1988). The species difference may arise from a rat-specific effect of diabetogenic agents on thyroid status resulting in hypothyroidism (Gotszche, 1983; Sundaresen *et al.*, 1984). It is known that thyroid hormones sensitize the heart to catecholamines (review: Collins *et al.*, 1989).

Cholinergic inhibition of cardiac adenylyl cyclase has been reported to be increased in streptozotocin-diabetic rats. This is associated with increased ratios of high- to low-affinity muscarinic cholinergic binding sites (Bergh *et al.*, 1984; Aronstam and Carrier, 1989). This may be interpreted as enhanced coupling between muscarinic receptors and G_i; however, no changes in $G_i\alpha$ have been documented to date. Remarkable as it may seem, the regional distribution of cardiac G-proteins in health or disease has not yet been explored in detail (but see Chapter 3) (Robishaw and Foster, 1989).

3.4 Nervous tissue

A number of behavioural, emotional and mood disorders are associated with diabetes mellitus (review: Lozovsky *et al.*, 1981). These effects are probably more related to long-term glucotoxicity rather than insulin withdrawal. Changes in central nervous system G-protein-coupled receptors and adenylyl cyclase activity have been reported (Lozovsky, 1981; Trulson and Himmel, 1983; Palmer *et al.*, 1983; Serri *et al.*, 1985; El-Refai and Chan, 1986; Laduron and Janssen, 1986; Welsh and Szabo, 1988; Shimizu *et al.*, 1990; Bitar and De Souza, 1990; Garris, 1990). Loss of G_i-mediated inhibition of adenylyl cyclase activity has also been reported in the corpus striatum (Abbracchio *et al.*, 1989) and in the retina (Hadjiconstantinou *et al.*, 1988) of chemically diabetic rats. The above findings coupled with the recent revelations concerning altered central nervous system inositol lipid metabolism in diabetes (Natarajan *et al.*, 1981; Fisher and Agranoff, 1987; Greene *et al.*, 1990) indicate a need for studies on central nervous system G-proteins in this disease. However, a careful regional examination of G-protein subunit immunoreactivity or mRNA levels has not been made.

3.5 Other tissues

In skeletal muscle, β-adrenergic, but not serotonergic or NaF, stimulation of adenylyl cyclase is reduced by diabetes (Garber, 1980). Decreases in both α_s and α_i mRNA levels have also been reported (Bushfield *et al.*, 1990c), but since the change in the G_s/G_i ratio is probably unaltered, the decrease in β-adrenergic responsiveness probably occurs at the receptor level.

Numerous alterations of G-protein-mediated signalling and receptors in lymphocytes and blood platelets from diabetic humans and animals have been reported (Waitzman *et al.*, 1977; Bonne *et al.*, 1979; Shepherd *et al.*, 1983; Serusclat *et al.*, 1983; Hamet *et al.*, 1985; Spalding *et al.*, 1986; Connell *et al.*, 1986; Noji *et al.*, 1986; Collier *et al.*, 1986; Abraham *et al.*, 1986; Thibonnier and Woloschak, 1988; Martin *et al.*, 1989; Jaschonek *et al.*, 1989; Senard *et al.*, 1990; Ishii *et al.*, 1990). However, no G-protein measurements have been reported in blood cells from diabetic animals to date.

G-protein-regulated cell signalling is also significantly altered in diabetic aortic smooth muscle (Palik *et al.*, 1982; MacLeod and McNeill, 1982; Cardwell and Webb, 1984; Scarborough and Carrier, 1984; Magoni *et al.*, 1989; Andersen *et al.*, 1985; Kamata *et al.*, 1989; Legan, 1989; Abiru *et al.*, 1990a,b; Lass *et al.*, 1989; Abebe *et al.*, 1990) but, here again, G-protein measurements are lacking. Of special interest would be studies on the aortic smooth muscle G-protein which couples α_1-adrenergic and other Ca^{2+}-mobilizing receptors to PIP_2

hydrolysis (presumably a member of the G_q family), since contractile responses are enhanced in the diabetic state (Abebe *et al.*, 1990).

4 GENERAL CONCLUSIONS

Although there is much evidence that cell-signalling systems involving G-proteins are deranged in several tissues in human diabetics and in animal models of types I and II diabetes, there is either little or no information, or considerable dispute, concerning changes in the levels of components of the systems. For example, chemical diabetes has been reported to increase, decrease or not change hepatic glucagon receptors or adenylyl cyclase and investigations of the changes in G_s and G_i have given contradictory results. Because of the toxic side effects of the agents used to induce chemical diabetes, genetically determined forms of diabetes in experimental animals are preferable models in which to study the disease. These also correspond more closely to the disease in humans, which cannot be studied in detail because of ethical considerations. Unfortunately, there have been very few studies of receptors, G-proteins and effector systems (adenylyl cyclase, PIP_2–phospholipase C) in animals that spontaneously exhibit syndromes similar to types I and II diabetes.

In the naturally occurring diabetes of the BB/Wor rat, hepatic adenylyl cyclase and α_s are increased substantially, without a change in α_i species. Similar findings with respect to the G-protein α subunits have been reported for humans with type II diabetes and in streptozotocin-diabetic rats by one group, but not another. In insulinopenic diabetic humans, the ratio between α_i and α_s in adipose tissue is unchanged, and in streptozotocin-diabetic rats, α_s, $\alpha_i 1$ and $\alpha_i 2$ are unaltered, although $\alpha_i 3$ is increased. There have been no reported changes in the levels of G-proteins in other tissues of animals with chemically induced or spontaneous diabetes.

In view of the situation described above, it is impossible to present a clear picture of the extent to which tissue responses to G-protein-linked hormones are altered in human or experimental diabetes. The situation with respect to receptors and G-proteins linked to adenylyl cyclase is very confused and there have been no systematic explorations of other systems, e.g. those linked to phosphoinositide phospholipase C, phosphatidylcholine phospholipase C, D and A_2, or ion channels. In view of the widespread incidence and serious consequences of diabetes, this lack of information is unfortunate. Studies to address these questions are clearly warranted.

REFERENCES

Abbracchio, M.P., DiLuca, M., DiGiulio, A.M., Cattabeni, F., Tenconi, B. and Gorio, A. (1989). *J. Neurosci. Res.* **24**, 517–523.

Abebe, W., Harris, K.H. and MacLeod, K.M. (1990). *J. Cardiovasc. Pharmacol.* **16**, 239–248.

Abiru, T., Kamata, K., Miyata, N. and Kasuya, Y. (1990a). *Can. J. Physiol. Pharmacol.* **68**, 882–888.

Abiru, T., Watanabe, Y., Kamata, K., Miyata, N. and Kasuya, Y. (1990). *Res. Commun. Chem. Pathol. Pharmacol.* **68**, 13–25.

Abraham, D.R., Hollingsworth, P.J., Smith, C.B., Jim, L., Zucker, L.B., Sobotka, P.A. and Vinik, A.I. (1986). *J. Clin. Endocrinol. Metab.* **63**, 906–912.

Allgayer, H., Bachmann, W. and Hepp, K.D. (1982) *Diabetologia,* **22**, 464–467.

Almira, E.C. and Misbin, R.I. (1989). *Metabolism* **38**, 102–103.

Andersen, E.B., Lindskov, H.O., Marving, J., Boesen, F., Beck-Nielsen, H. and Hesse, B. (1985). *Dan. Med. Bull.* **32**, 194–196.

Appel, M.C., Like, A.A., Rossini, A.A., Camp, D.B. and Miller, T.B., Jr (1981). *Am. J. Physiol.* **240**, E38–E87.

Arner, P. (1988). *Diabetes/Metabolism Rev.* **4**, 507–515.

Arner, P. and Östman, J. (1976). *Diabetologia* **123**, 593–599.

Arner, P., Engfeldt, P. and Östman, J. (1979). *Metabolism* **28**, 198–209.

Arner, P., Engfeldt, P., Wennlund, A. and Östman, J. (1981). *Horm. Metab. Res.* **13**, 272–276.

Aronstam, R.S. and Carrier, G.O. (1989). *Diabetes* **38**, 1611–1616.

Atkins, F.L., Dowell, T. and Love, S. (1985) *J. Cardiovasc. Pharmacol.* **7**, 66–70.

Bell, G. I., Kayano, T., Buse, J. B., Burant, C. F., Takeda, J., Lin, D., Fukumoto, H. and Seino, J. (1990). *Diabetes Care* **13**, 198–208.

Bergh, C.-H., Ransnas, L., Hjalmarson, A., Waldenstrom, A. and Jackson, B. (1984). *Acta Pharmacol. Toxicol.* **55**, 373–379.

Bergh, C.-H., Hjalmarson, A., Sjogren, K.-G. and Jacobsson, B. (1988). *Horm. Metab. Res.* **20**, 381–386.

Berlin, I., Grimaldi, A., Bosquet, F. and Puech, A.J. (1986). *J. Clin. Endocrinol. Metab.* **63**, 262–265.

Bhathena, S.J., Voyles, N.R., Smith, S. and Recant, L. (1978). *J. Clin. Invest.* **61**, 1488–1497.

Bitar, M.S. and De Souza, E.B. (1990). *J. Pharmacol. Exp. Ther.* **254**, 781–785.

Bitar, M.S., Koulu, M., Rapoport, S.I. and Linnoila, M. (1987). *Biochem. Pharmacol.* **36**, 1011–1016.

Blackmore, P.F. and Exton, J.H. (1986). *J. Biol. Chem.* **261**, 11056–11063.

Bondy, P.K. and Felig, P. (1974) In: *Duncan's Diseases of Metabolism,* 7th edn (Bondy, P.K. and Rosenberg, L.E., eds), pp. 221–340, W.B. Saunders Co., Philadelphia.

Bonne, C., Romquin, N. and Regnault, F. (1979). *Bibl. Anat.* **18**, 98–103.

Burgess, G.M., Bird, G.St.J., Obie, J.F. and Putney, J.W. Jr (1991). *J. Biol. Chem.* **266**, 4772–4781.

Burns, T.W., Langley, P.E. and Robison, G.A. (1972). *Adv. Cyclic Nucleotide Res.* **1**, 63–85.

Bushfield, M., Griffiths, S.L., Murphy, G.J., Pyne, N.J., Knowler, J.T., Milligan, G., Parker, P.J., Mollner, S. and Houslay, M.D. (1990a) *Biochem. J.* **271**, 365–372.

Bushfield, M., Murphy, G.J., Lavan, B.E., Parker, P.J., Hruby, V.J., Milligan, G. and Houslay, M.D. (1990b). *Biochem. J.* **268**, 449–457.

Bushfield, M., Griffiths, S.L., Strassheim, D., Tang, E., Shakur, Y., Lavan, B. and Houslay, M.D. (1990c). *Biochem. Soc. Symp.* **56**, 137–154.

Cardwell, R.J. and Webb, R.C. (1984). *Clin. Physiol.* **4**, 509–517.

Caro, J.F., Sinha, M.K. and Dohm, G.L. (1990). In: *Obesity: Towards a Molecular Approach* (Bray, G.A., Ricquier, D. and Spiegelman, B.M., eds), pp. 203–217. Alan R. Liss, Inc. New York.

Caro, J.F., Raju, M.S., Exton, J.H., Thakkar, J.K., Roy, L., Hale, J.C. and Pories, W.J. (1991). *J. Biol. Chem.* (suppl. **15B**, 13).

Chamras, H., Fouchereau-Peron, M. and Rosselin, G. (1980). *Diabetologia* **19**, 74–80.

Chan, T.M., Young, K.M., Hutson, N.J., Brumley, F.T. and Exton, J.H. (1975). *Am. J. Physiol.* **229**, 1702–1712.

Chang, A.Y., Noble, R.E. and Wyse, B.M. (1977). *Diabetes* **26**, 1063–1071.

Chatelain, P., Gillet, L., Waelbroeck, M., Camus, J.C., Robberecht, P. and Christophe, J. (1983). *Horm. Metab. Res.* **15**, 620–622.

Chiappe de Cingolani, G.E. (1983). *Arch. Int. Physiol. Biochim.* **91**, 1–8.

Chiappe de Cingolani, G.E. (1986a). *Acta Physiol. Pharmacol. Latinoam* **36**, 39–46.

Chiappe de Cingolani, G.E. (1986b). *Diabetes* **35**, 1229–1232.

Christensen, N.J. (1974) *Diabetes* **23**, 1–8.

Christophe, J., Waelbroeck, M., Chatelain, P. and Robberecht, P. (1984). *Peptides* **5**, 341–353.

Codina, J., Olate, J., Abramowitz, J., Mattera, R., Cook, R.G. and Birnbaumer, L. (1988). *J. Biol. Chem.* **263**, 6746–6750.

Collier, A., Tymkewycz, P., Armstrong, R., Young, R.J., Jones, R.L. and Clarke, B.F. (1986). *Diabetologia* **29**, 417–474.

Collins, S., Bolanowski, M.A., Caron, M.G. and Lefkowitz, R.J. (1989). *Annu. Rev. Physiol.* **51**, 203–215.

Connell, J.M.C., Ding, Y.-A., Fisher, B.M., Frier, B.M. and Semple, P.F. (1986). *Clin. Sci.* **71**, 217–220.

Crane, J.K., Campanile, C.P. and Garrison, J.C. (1982). *J. Biol. Chem.* **257**, 4951–4958.

Cryer, P.E. (1980). *N. Engl. J. Med.* **303**, 436–444.

deGaetano, G. (1981). *Clinics Haematol.* **10**, 297–326.

Dighe, R.R., Rojas, F.J., Birnbaumer, L. and Garber, A.J. (1984). *J. Clin. Invest.* **73**, 1013–1023.

Eisen, H.J. and Goodman, H.M. (1969). *Notes and Comments* **84**, 414–416.

El-Hage, A.N., Herman, E.H., Jordan, A.W. and Ferrans, V.J. (1985). *J. Mol. Cell. Cardiol.* **17**, 361–369.

El-Refai, M.F. and Chan, T.M. (1986). *Biochim. Biophys. Acta* **880**, 16–25.

Engfeldt, P., Arner, P., Bolinder, J. and Östman, J. (1982). *J. Clin. Endocrinol. Metab.* **55**, 983–988.

Exton, J.H. (1980). *Adv. Cyclic Nucleotide Res.* **12**, 319–327.

Exton, J.H. (1987). *Diabetes Metab. Rev.* **3**, 163–183.

Exton, J.H. (1991) *Diabetes* **40**, 521–526.

Exton, J.H., Assimacopoulos-Jeannet, F.P., Blackmore, P.F., Cherrington, A.D. and Chan, T.M. (1978). *Adv. Cyclic Nucleotide Res.* **9**, 441–452.

Fain, J.N. and Garcia-Sainz, J.A. (1983). *J. Lipid Res.* **24**, 945–966.

Fein, F.S. and Scheuer, J. (1990). In: *Ellenberg and Rifkin's Diabetes Mellitus, Theory and Practice* (Rifkin, H. and Porte, D. Jr, eds), pp. 812–823. Elsevier Science Publishing Co. Inc., New York.

Fisher, S.K. and Agranoff, B.W. (1987). *J. Neurochem.* **48**, 999–1017.

Fujita, T., Yui, R. and Kusomoto, Y. (1982). *Biomed. Res.* **3**, 429–436.

Garber, A.J. (1980). *J. Clin. Invest.* **65**, 478–487.

Garris, D.R. (1990). *Dev. Brain Res.* **51**, 161–166.

Gawler, D., Milligan, G., Spiegel, A.M., Unson, C.G. and Houslay, M.D. (1987). *Nature* **327**, 229–232.

Ginsbert-Fellner, F. and Knittle, J.L. (1973). *Diabetes* **22**, 528–536.

Goldberg, N.D., Dietz, S.B. and O'Toole, A.G. (1969). *J. Biol. Chem.* **244**, 4458–4466.

Goodman, H.M. (1970). *Endocrinology* **86**, 1064–1074.

Gotzche, O. (1983). *Diabetes* **32**, 1110–1116.

Granner, D. and Pilkis, S. (1990). *J. Biol. Chem.* **265**, 10173–10176.

Greene, D.A., Sima, A.A.F., Albers, J.W. and Pfeifer, M.A. (1990). In: *Ellenberg and Rifkin's Diabetes Mellitus, Theory and Practice* (Rifkin, H. and Porte, D. Jr, eds), pp. 710–755. Elsevier Science Publishing Co. Inc., New York.

Griffiths, S.L. and Houslay, M.D. (1989). *Biochem. Soc. Trans.* **18**, 475–476.

Griffiths, S.L., Knowler, J.T. and Houslay, M.D. (1990). *Eur. J. Biochem.* **193**, 367–374.

Hadjiconstantinou, M., Qu, Z.-X., Moroi-Fetters, S.E. and Neff, N.H. (1988). *Eur. J. Pharmacol.* **149**, 193–194.

Hamet, P., Skuherska, R., Pang, S.C. and Tremblay, J. (1985). *Hypertension* **7**, II135–II142.

Hansen, A.P. (1971). *J. Clin. Invest.* **50**, 1806–1811.

Hayford, J.T., Danny, M.M., Hendrix, J.A. and Thompson, R.G. (1980). *Diabetes* **29**, 391–398.

Herberg, L. and Coleman, D.L. (1977). *Metabolism* **26**, 59–99.

Heyworth, C.M., Whetton, A.D., Wong, S., Martin, B.R. and Houslay, M.D. (1985). *Biochem. J.* **228**, 593–603.

Ingebretsen, C.G., Hawelu-Johnson, C. and Ingebretsen, W.R. Jr (1983). *J. Cardiovasc. Pharmacol.* **5**, 454–461.

Ingebretsen, W.R., Peralta, C., Monsher, M., Wagner, L.K. and Ingebretsen, C.G. (1981). *Am. J. Physiol.* **240**, H375–H382.

Ishii, H., Umeda, F., Hashimoto, T. and Nawata, H. (1990). *Diabetes* **39**, 1561–1568.

Israelsson, B. and Tengrup, I. (1980). *Experientia* **36**, 257–258.

Jackson, C.V., McGrath, G.M. and McNeill, J.H. (1986). *Can. J. Physiol. Pharmacol.* **64**, 145–151.

Jard, S., Cantau, B. and Jakobs, K.H. (1981). *J. Biol. Chem.* **256**, 2603–2606.

Jaschonek, K., Faul, C., Weisenberger, H., Krönert, K., Schröder, H. and Renn, W. (1989). *Thrombosis Haemostasis* **61**, 535–536.

Jefferson, L.S., Exton, J.H., Butcher, R.W., Sutherland, E.W. and Park, C.R. (1968). *J. Biol. Chem.* **243**, 1031–1038.

Johnson, D.G., Goebel, C.U., Hruby, V.J., Bregman, M.D. and Trivedi, D. (1982). *Science* **215**, 1115–1116.

Junod, A., Lambert, A.E., Stauffacher, W. and Renold, A.E. (1969). *J. Clin. Invest.* **48**, 2129–2139.

Kamata, K., Miyata, N. and Kasuya, Y. (1989). *J. Pharmacol. Exp. Ther.* **249**, 890–894.

Kampa, I.S. and Frascella, D.W. (1977). *Horm. Metab. Res.* **9**, 282–285.

Kampa, I.S., Frascella, D.W., Hundertmark, S.M., Rosenberg, J.M. and Reid, V. (1981). *Horm. Metab. Res.* **13**, 122.

Kashiwagi, A., Nishio, Y., Saeki, Y., Kida, Y., Kodama, M. and Shigeta, Y. (1989). *Am. J. Physiol.* **257**, E127–E132.

Kenno, K.A. and Severson, D.L. (1985). *Am. J. Physiol.* **249**, H1024–H1030.

Kupiecki, F.P. and Adams, L.D. (1974). *Diabetologia* **10**, 633–637.

Lacasa, D., Agli, B. and Gindicelli, Y. (1983). *Eur. J. Biochem.* **130**, 457–464.

Laduron, P.M. and Janssen, P.F.M. (1986). *Brain Res.* **380**, 359–362.

Lass, P., Knudsen, G.M., Pedersen, E.V. and Barry, D.I. (1989). *Pharmacol. Toxicol.* **65**, 318–320.

Laube, H., Fussgänger, R.D., Maier, V. and Pfieffer, E.F. (1973). *Diabetologia* **9**, 400–402.

Lefkowitz, R.J., Hoffmann, B.B. and Taylor, P. (1990). In: *Goodman and Gilman's, The Pharmacological Basis of Therapeutics*, 8th Edition (Goodman, A.G., Rall, T.W., Nies, A.S. and Taylor, P., eds). Pergamon Press, New York, N.Y., pp. 84–121.

Legan, E. (1989). *Life. Sci.* **45**, 371–378.

Leibel, R.L. and Hirsch, J. (1987). *J. Clin. Endocrin. Metab.* **64**, 1205–1210.

Lenzen, S. and Panten, U. (1988). *Diabetologia* **31**, 337–342.

Levilliers, J., Pairault, J., Lecot, F., Tournemole, A. and Laudat, M.-H. (1978). *Eur. J. Biochem.* **88**, 323–330.

Lockwood, D.H., Hamilton, C.L. and Livingston, J. N. (1979). *Endocrinology* **104**, 76–81.

Lozovsky, D., Saller, C.F. and Kopin, I.J. (1981). *Science* **214**, 1031–1033.

Lynch, C.J., Blackmore, P.F., Johnson, E.H., Wange, R.L., Krone, P.K. and Exton, J.H. (1989). *J. Clin. Invest.* **83**, 2050–2062.

MacLeod, K.M. and McNeill, J.H. (1982). *Proc. West. Pharmacol. Soc.* **25**, 245–247.

Madar, Z. (1984). *Horm. Metab. Res.* **16**, 59–62.

Magoni, M.S., Kobayashi, H., Trezzi, E., Catapano, A., Spano, P.F. and Trabucchi, M. (1984). *Life. Sci.* **34**, 1095–1100.

Martin, S., Kolb-Bachofen, V., Kiesel, U. and Kolb, H. (1989). *Diabetologia* **32**, 140–142.

Mattera, R., Yatani, A., Kirsch, G.E., Graf, R., Okabe, K., Olate, J., Codina, J., Brown, A.M. and Birnbaumer, L. (1989). *J. Biol. Chem.* **264**, 465–471.

Mayor, P. and Calle, C. (1988). *Endocrinol. Japon.* **35**, 207–215.

McKechnie, J.K., Leary, W.P., Noakes, T.D., Kallmeyer, J.C., MacSearraigh, E.T.M. and Oliver, L.R. (1979). *S. Afr. Med. J.* **56**, 256–261.

Menahan, L.A., Chaudri, S.N., Weber, H.E. and Shipp, J.C. (1977). *Horm. Metab. Res.* **9**, 527–528.

Michel, A., Cros, G.H., McNeill, J.H. and Serrano, J.J. (1985). *Life. Sci.* **37**, 2067–2075.

Miller, T.B. Jr (1978). *Am. J. Physiol.* **234**, E13–E19.

Miller, T.B. Jr (1981). *J. Biol. Chem.* **256**, 1748–1753.

Miller, T.B. Jr (1983). *Am. J. Physiol.* **245**, E379–E383.

Miller, T.B. Jr (1984). *Am. J. Physiol.* **246**, E134–E140.

Miller, T.B., Jr, Vicalvi, J.J. and Garnache, A.K. (1981). *Am. J. Physiol.* **240**, E539–E543.

Murphy, G.J., Gawler, D.J., Milligan, G., Wakelam, M.J.O., Pyne, N.J. and Houslay, M.D. (1989). *Biochem. J.* **259**, 191–197.

Nakhooda, A.F., Like, A.A., Chappel, C.I., Murray, F.T. and Marliss, E.B. (1977). *Diabetes* **26**, 100–112.

Natarajan, V., Dyck, P.J. and Schmid, H.H.O. (1981). *J. Neurochem.* **36**, 413–419.

Nishio, Y., Kashiwagi, A., Kida, Y., Kodama, M., Abe, N., Saeki, Y. and Shigeta, Y. (1988). *Diabetes* **37**, 1181–1187.

Noji, T., Tashiro, M., Nagashima, K., Suzuki, S. and Kuroume, T. (1986). *Horm. Metab. Res.* **18**, 604–606.

Ohisalo, J.J., Vikman, H.-L., Ranta, S., Houslay, M.D. and Milligan, G. (1989). *Biochem. J.* **264**, 289–292.

Opie, L.H., Tansey, M.J. and Kennelly, B.M. (1979). *S. Afr. Med. J.* **56**, 207–211.

Palik, I., Koltai, M.Z., Wagner, M., Kolonics, I. and Pogatsa, G. (1982) *Experientia* **38**, 934–935.

Palmer, G.C., Wilson, G.L. and Chronister, R.B. (1983). *Life Sci.* **32**, 365–374.

Pilkis, S.J., Exton, J.H., Johnson, R.A. and Park, C.R. (1974). *Biochim. Biophys. Acta* **343**, 250–267.

Portha, B., Chamras, H., Broer, Y., Picon, L. and Rosselin, G. (1983). *Mol. Cell. Endocrinol.* **32**, 13–26.

Ramanadham, S. and Tenner, T.E. Jr (1983). *Pharmacology* **27**, 130–139.

Ramanadham, S. and Tenner, T.E. Jr (1986). *Diabetologia* **29**, 741–748.
Ramanadham, S. and Tenner, T.E. Jr (1987). *Eur. J. Pharmacol.* **136**, 377–389.
Ramanadham, S., McNeill, J.H. and Tenner, T.E. Jr (1986). *Proc. West. Pharmacol. Soc.* **29**, 23–25.
Robishaw, J.D. and Foster, K.D. (1989). *Annu. Rev. Physiol.* **51**, 229–244.
Rothenberg, P.L. and Kahn, C.R. (1988). *J. Biol. Chem.* **263**, 15546–15552.
Saggerson, E.D. (1989). *Biochem. Soc. Trans.* **17**, 47–49.
Sato, N., Irie, M., Kajinuma, H. and Suzuki, K. (1989). *Acta Endocrinol.* **121**, 705–713.
Savarese, J.J. and Berkowitz, B.A. (1979). *Life. Sci.* **25**, 2075–2078.
Scarborough, N.L. and Carrier, G.O. (1984). *J. Pharmacol. Exp. Ther.* **231**, 603–609.
Schimmel, R.J. (1976). *Biochim. Biophys. Acta* **451**, 363–371.
Seitz, H.J., Lüth, W. and Tarnowski, W. (1979). *Arch. Biochem. Biophys.* **195**, 385–391.
Senard, J.M., Barbe, P., Estan, L., Guimbaud, R., Louvet, J.P., Berlan, M., Tran, M.A. and Montastruc, J.L. (1990). *J. Clin. Endocrinol. Metab.* **71**, 311–317.
Senft, G., Schultz, G., Munske, K. and Hoffmann, M. (1968). *Diabetologia* **4**, 322–329.
Sengupta, K., Long, K.J. and Allen, D.O. (1981). *J. Pharmacol. Exp. Ther.* **216**, 15–19.
Serri, O., Renier, G. and Somma, M. (1985). *Horm. Res.* **21**, 95–101.
Serusclat, P., Rosen, S.G., Smith, E.B., Shah, S.D., Clutter, W.E. and Cryer, P.E. (1983). *Diabetes* **32**, 825–829.
Shafrir, E. (1990). In: *Ellenberg and Rifkin's Diabetes Mellitus: Theory and Practice* (Rifkin, H. and Porte, D. Jr, eds), pp. 291–340. Elsevier Publishing Co. Inc., New York.
Shamoon, H., Hendler, R. and Sherwin, R.S. (1980). *Diabetes* **29**, 284–291.
Shepherd, G.L., Lewis, P.J., Blair, I.A., deMey, C. and MacDermot, J. (1983). *Br. J. Clin. Pharmacol.* **15**, 77–81.
Shikama, H. and Ui, M. (1976). *Biochim. Biophys. Acta* **444**, 461–471.
Shimizu, H., Shimomura, Y., Takahashi, M., Kobayashi, I. and Kobayashi, S. (1990). *Exp. Clin. Endocrinol.* **95**, 263–266.
Smith, C.I., Pierce, G.N. and Dhalla, N.S. (1984). *Life Sci.* **34**, 1223–1230.
Solomon, S.S., Heckemeyer, C.M., Barker, J.A. and Duckworth, W.C. (1985). *Endocrinology* **117**, 1350–1354.
Solomon, S.S., Steiner, M.S., Sanders, L. and Palazzolo, M.R. (1986). *Endocrinology* **119**, 1839–1844.
Solomon, S.S., Sibley, S.D. and Cunningham, T.M. (1990). *Endocrinology* **127**, 1544–1546.
Soman, V. and Felig, P. (1978). *J. Clin. Invest.* **61**, 552–560.
Spalding, R.M., Ward, W.K., Malpass, T.W., Stratton, J.R., Halter, J.B., Porte, D. Jr and Pfeifer, M.A. (1986). *Diabetes Care* **9**, 276–278.
Srikant, C.B., Freeman, D., McCorkle, K. and Unger, R.H. (1977). *J. Biol. Chem.* **252**, 7434–7436.

Srivastava, A.K. and Anand-Srivastava, M.B. (1985). *Biochem. Pharmacol.* **34**, 2013–2017.

Strassheim, D., Milligan, G. and Houslay, M.D. (1990). *Biochem. J.* **266**, 521–526.

Studer, R.K. and Ganas, L. (1989). *Endocrinology* **125**, 2421–2433.

Sundaresan, P.R., Sharma, V.K., Gingold, I. and Banerjee, S.P. (1984). *Endocrinology* **114**, 1358–1363.

Tamborlane, W.V., Sherwin, R.S., Koivisto, V., Hendler, R., Genel, M. and Felig, P. (1979). *Diabetes* **28**, 785–788.

Taylor, S.J., Smith, J.A. and Exton, J.H. (1990). *J. Biol. Chem.* **265**, 17150–17156.

Taylor, S.J., Chae, H.Z., Rhee, S.G. and Exton, J.H. (1991). *Nature* **350**, 516–518.

Thakker, J.K., DiMarchi, R., MacDonald, K. and Caro, J.F. (1989). *J. Biol. Chem.* **264**, 7169–7175.

Thibonnier, M. and Woloschak, M. (1988). *Proc. Soc. Exp. Biol. Med.* **188**, 149–152.

Topper, E., Gertner, S., Amiel, S., Press, M., Genel, M. and Tamborlane, W.V. (1985). *Ped. Res.* **19**, 1985.

Trulson, M.E. and Himmel, C.D. (1983). *J. Neurochem.* **40**, 1456–1459.

Unger, R.H. and Orci, L. (1976). *Physiol. Rev.* **56**, 778–826.

Unger, R.H. and Orci, L. (1981). *New Engl. J. Med.* **304**, 1575–1580.

Wager, J., Fredholm, B.B., Lunell, N.-O. and Persson, B. (1981). *Br. J. Obstet. Gynaecol.* **88**, 352–361.

Wagle, S.R., Ingebretsen, W.R. Jr and Sampson, L. (1975). *Diabetologia* **11**, 411–417.

Wahrenberg, H., Lönnqvist, F., Engfeldt, P. and Arner, P. (1989). *Diabetes* **38**, 524–533.

Wahrenberg, H., Arner, P., Adamsson, U., Lins, P.E., Juhlin-Dannfelt, A. and Östman, J. (1990). *J. Intern. Med.* **227**, 309–316.

Waitzman, M.B. (1979). *Metabolism* **28**, 401–406.

Waitzman, M.B., Colley, A.M. and Nardelli-Olkowska, K. (1977). *Diabetes* **25**, 510–519.

Wald, M., Enri, S.B. and Sterin-Borda, L. (1988). *Can. J. Physiol. Pharmacol.* **66**, 1154–1160.

Wange, R.W., Smrcka, A.V., Sternweis, P.C. and Exton, J.H. (1991). *J. Biol. Chem.,* **266**, 11409–11412.

Wasserman, D.H., Lickley, H.L.A. and Vranic, M. (1985). *J. Appl. Physiol.* **59**, 1272–1281.

Weiss, R.B. (1982). *Cancer Treatment Rep.* **66**, 427–438.

Welsh, J.B. and Szabo, M. (1988). *Endocrinology* **123**, 2230–2234.

Williams, R.S., Schaible, T.F., Scheuer, J. and Kennedy, R. (1983). *Diabetes* **32**, 881–886.

Wong, Y.H., Federman, A., Pace, A.M., Zachary, I., Evans, T., Pouyssegur, J. and Bourne, H.R. (1991). *Nature* **351**, 63–65.

Yamashita, K., Yamashita, S., Yasuda, H., Oka, Y. and Ogata, E. (1980). *Diabetes* **29**, 188–192.

Zaagsma, J. and Nahorski, S.R. (1990). *TIPS* **11**, 3–7.

Zanoboni, A. and Zanoboni-Muciaccia, W. (1974). *Specialia* **31**, 473–474.

Zapf, J., Feuerlein, D., Waldvogel, M. and Froesch, E.R. (1975). *Diabetologia* **11**, 509–516.

Zapf, J., Waldvogel, M., Zumstein, P. and Froesch, E.R. (1978). *FEBS Lett.* **94**, 43–46.

Zola, B.E., Miller B., Stiles, G.L., Rao, P.S., Sonnenblick, E.H. and Fein, F.S. (1988). *Am. J. Physiol.* **255**, E636–E641.

Zumstein, P., Zapf, J., Waldvogel, M. and Froesch, E.F. (1980). *Eur. J. Biochem.* **105**, 187–194.

CHAPTER FIVE

Adrenal Dysfunction and G-protein-mediated Pathways

JOHN R. HADCOCK* and CRAIG C. MALBON
*Diabetes and Metabolic Diseases Research Program,
Department of Pharmacology, School of Medicine,
Health Sciences Center, SUNY, Stony Brook, New
York 11794-8651, USA*
* *Present address: Molecular and Cellular Biology
Group, American Cynamid Company, Agricultural
Research Division, Princeton, New Jersey
08543-0400, USA*

1 INTRODUCTION

The hypothalamic–pituitary–adrenal axis regulates glucocorticoid secretion by a complex and coordinated feedback system. Secretion of corticotropin-releasing hormone (CRH) from the hypothalamus promotes secretion of adrenocorticotropin (ACTH) by the pituitary. CRH activates specific receptors on pituitary gland corticotropes and shows significant diurnal variation (Labrie *et al.*, 1976). CRH receptors, coupled to adenylyl cyclase via G_s, elevate intracellular cAMP and activate cAMP-dependent protein kinase (protein kinase A) (Labrie *et al.*, 1982). The increased circulating ACTH levels, in turn, activate specific receptors in the adrenal cortex. These receptors for ACTH also stimulate adenylyl cyclase via G_s, activate protein kinase A and increase the rate of steroidogenesis. The diurnal variation in CRH secretion accounts for the large fluctuations in circulating plasma cortisol levels. A negative feedback regulation by glucocorticoids is a major mechanism for maintaining the regulation of secretion of glucocorticoids by ACTH. Excess glucocorticoids decrease the secretion of both CRH and ACTH.

Both glucocorticoid excess and deficiency can have profound effects on G-protein-mediated pathways (Haynes, 1990). Many tissue-specific

G-Proteins
ISBN 0-12-497515-1

effects of glucocorticoids on transmembrane signalling have been observed. Via 'permissive effects', glucocorticoids regulate transmembrane signalling and the metabolic pathways controlled by a variety of hormones, neurotransmitters and autacoids (Malbon *et al.*, 1988). Glucocorticoid deficiency, known as 'Addison's disease', is deleterious. Impairments of motor activity, of the cardiovascular system and of several metabolic pathways regulated via receptor–G-protein–effector motifs have been described. Prominent among the deleterious effects of glucocorticoid insufficiency is the greatly enhanced sensitivity to stress which, in the extreme, can cause death. Fortunately, replacement therapy with glucocorticoids readily normalizes many of these physiological processes. Glucocorticoid insufficiency generally results from a decrease in the function of the hypothalamic–pituitary axis. Reductions in ACTH secretion, for example, will cause glucocorticoid insufficiency. Adrenal corticosteroid excess (Cushing's syndrome), resulting from either a disease process or pharmacological doses of glucocorticoids, can have profound effects on metabolic pathways. Hyperglycaemia, inhibition of protein synthesis and redistribution of fat depots are commonly observed in chronic hyperfunction of the adrenals.

Most of our knowledge of the physiological, biochemical and molecular effects of glucocorticoids on G-protein-mediated pathways is derived from studies of animal models and cultured cell lines. The focus of this chapter is a description of tissue-specific changes induced by glucocorticoids, directed primarily at the levels of receptors and G-proteins, and to a lesser extent upon effectors.

2 ADIPOSE TISSUE

Stimulation of β-adrenergic receptors by catecholamines and stimulation of other G-protein-linked receptors by their agonist ligands regulate the lipolytic response of white adipose tissue (Lands *et al.*, 1967; Braun and Hechter, 1970). For β-adrenergic agonists, at least three target receptor subtypes exist, β_1, β_2 and β_3 (Lands *et al.*, 1967; Emorine *et al.*, 1989). Some controversy exists as to which of the adrenergic receptor subtypes (β_1 or β_3) mediates catecholamine-stimulated lipolysis in rat white adipose tissue (Bahouth and Malbon, 1988; Bojanic *et al.*, 1985; Emorine *et al.*, 1989). Based upon the purification of the rat adipocyte β-adrenergic receptor, radioligand binding and the pharmacology of hormone-stimulated lipolysis, a β_1-adrenergic receptor subtype, albeit 'atypical', appears to be the predominant species expressed in rat adipocytes (Lands *et al.*, 1967; Moxham and Malbon, 1985; Bahouth and Malbon, 1988). mRNA tissue distribution has identified β_3-adrenergic receptor mRNA in mouse white adipose tissue (Emorine *et al.*, 1989), although it has not yet been identified for adipose tissue of either rat or human (Emorine *et al.*, 1989). Receptor-mediated elevation of intracellular cAMP levels and activation of protein kinase A promotes phosphorylation and activation of a hormone-sensitive lipase. This lipase catalyses the hydrolysis of stored triglycerides, generating fatty acids and glycerol to fuel generation of ATP. Adrenalectomy markedly impairs both catecholamine-stimulated cAMP accumulation and lipolysis in adipose tissue (Ros *et al.*, 1989a; Allen and Beck, 1972; Figure 1). The impaired lipolytic response to catecholamines observed in adrenalectomized animals can be partially reversed by the treatment of animals with glucocorticoids. The underlying mechanism(s) responsible for the attenuated response in adrenalectomy appears to be complex. Elements that may be regulated via adrenalectomy include not only receptors, but G-proteins and adenylyl cyclase as well. Other elements distal to the transmembrane signalling machinery may also be targets for regulation by glucocorticoids.

At the level of receptor, adipocyte β-adrenergic receptor levels are not significantly altered in adrenalectomized female rats (Ros *et al.*, 1989a). Thotakura *et al.* (1982), however, observed a 35% decline in β-adrenergic receptor number in adipocytes isolated from adrenalectomized male rats. Treating adrenalectomized animals with the synthetic glucocorticoid dexamethasone increases the expression of β-adrenergic receptors in adipocytes and other tissues (Ros *et al.*, 1989a). Likewise, in primary cultures of fat cells from rats treated with dexamethasone, β-adrenergic receptor levels are elevated (Lacasa *et al.*, 1988). Adrenaline-stimulated cAMP accumulation and lipolysis are both impaired in adrenalectomized rats (Figure 1). Treating adrenalectomized animals partially restores adrenaline-stimulated lipolysis, but not cAMP accumulation (Ros *et al.*, 1989a; Figure 1). An increase in β-adrenergic receptor number during dexamethasone treatment explains, in part, the restoration of catecholamine-stimulated lipolysis observed in the steroid-treated, adrenalectomized animals. Interestingly, the responsiveness to catecholamines in adipocytes from adrenalectomized rats can also be enhanced by treating cells with adenosine deaminase to remove endogenous adenosine (Thotakura *et al.*,

1982). These results suggest that the inhibitory pathway of adenylyl cyclase activated by adenosine may also play a role in modulating cellular responsiveness to catecholamines.

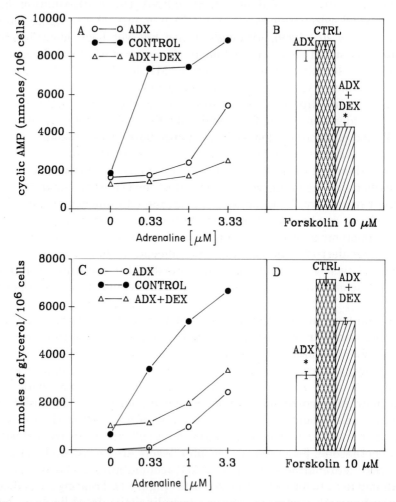

Figure 1. Adrenaline- and forskolin-stimulated cAMP accumulation and lipolysis: effects of adrenalectomy. Adipocytes from adrenalectomized (ADX), control (CTRL), and adrenalectomized, dexamethasone-treated (ADX + DEX) rats were isolated and incubated with either adrenaline (panels A and C) or forskolin (panels B and D). Intracellular cAMP accumulation (panels A and B) and glycerol release (panels C and D) were then measured. *Significantly different from control ($P < 0.05$). Reprinted with permission from Ros *et al.* (1989a).

$G_s\alpha$, $G_i\alpha1,2,3$ and at least one form of $G_o\alpha$ are expressed in rat adipocytes (Rapiejko *et al.*, 1990; Green *et al.*, 1990; Longabaugh *et al.*, 1990). In adrenalectomized rats the steady-state levels of $G_s\alpha$ in adipose tissue decline by 30–40% as compared to control animals. Treating adrenalectomized rats with dexamethasone not only restores but actually increases the levels of $G_s\alpha$ above that observed for control animals. The levels of the mRNA encoding several of these G-protein subunits have been quantified recently (Rapiejko *et al.*, 1990). Interestingly, the steady-state levels of $G_s\alpha$ mRNA in rat adipocytes appear to be unaltered by either adrenalectomy or by glucocorticoid treatment (Ros *et al.*, 1989b), suggesting a post-translational mode of regulation. In rat adipocytes, the steady-state expression of β1 and β2 are regulated also by adrenalectomy as well as by glucocorticoids. The steady-state levels of Gβ1 and Gβ2 decline by 50% in adipocytes from adrenalectomized rats (Ros *et al.*, 1989a,b). The decline of the cellular content of Gβ1 and Gβ2 can be ascribed to a prominent decline in the mRNA levels for the two β subunits (Ros *et al.*, 1989b).

By changing the cellular content of G-protein subunits, glucocorticoids alter the ratio of α to β subunits, influencing hormone-stimulated cAMP accumulation and lipolysis. Alteration in the ratio of α to β subunits is likely to play a key role in dictating the responsiveness of the hormone-stimulated adenylyl cyclase both in adipocytes isolated from adrenalectomized rats and perhaps in other target tissues (Bokoch, 1987; Gilman, 1987; Watkins *et al.*, 1989). In contrast to the alterations observed for receptors and for G-proteins, the activity of the catalyst adenylyl cyclase, as measured by forskolin-stimulated cAMP accumulation, does not appear to be significantly altered in adipocytes after adrenalectomy (Ros *et al.*, 1989a). Even though forskolin-stimulated cAMP accumulation is normal in adipocytes from adrenalectomized rats, forskolin-stimulated lipolysis is only 50% of normal (Ros *et al.*, 1989; Figure 1). Thus it appears that the decline in hormone-stimulated lipolysis is partially dependent on the decline in both the expression of receptors and the G-proteins to which they are coupled. However, other component(s) distal to the transmembrane signalling apparatus appear to be impaired by adrenalectomy. Protein kinase A or the hormone-sensitive lipase activities provide likely targets. Understanding glucocorticoid action in adipose tissue will require a more detailed analysis of the pathway extending from the G-protein-linked receptor to the lipase.

3 BRAIN

Glucocorticoids have pronounced behavioural and physiological effects in the brain. Administration of glucocorticoids can lead to mood swings and to hallucinations (see references in McEwen *et al.* (1986) and Meyer (1985)). Some of these changes in behavioural patterns have been ascribed to changes in G-protein-mediated pathways resulting from glucocorticoid imbalances (McEwen *et al.*, 1986; Meyer, 1985). Not only are some pathways of neurotransmitter synthesis altered, but several G-protein-mediated pathways are impaired functionally. The catecholamine-stimulated adenylyl cyclase in rat cortex is mediated by both β-adrenergic and α-adrenergic receptors. While adrenergic stimulation of adenylyl cyclase is mediated predominantly via the β-adrenergic receptor pathway, potentiation by an α-adrenergic receptor pathway has been demonstrated in several tissues (Exton *et al.*, 1972; Johnson *et al.*, 1986; Perkins and Moore, 1973). The α-adrenergic receptor pathway also potentiates adenosine-stimulated cAMP accumulation in the rat frontal cortex (Duman *et al.*, 1989). This potentiation of $G_s\alpha$-mediated pathways is probably mediated by α_1-rather than α_2-adrenergic receptors. α_1-Adrenergic receptors are well-known stimulators of phospholipase C (PLC) activity. PLC activation generates two second messengers, inositol 1,4,5-trisphosphate (IP_3) and diacylglycerol (DAG). Increases in IP_3 promote intracellular Ca^{2+} mobilization and increases in DAG activate protein kinase C. α_2-Adrenergic receptors, in contrast, mediate inhibition of adenylyl cyclase in many tissues (Limbird, 1988). The noradrenaline-mediated potentiation of the G_s-mediated pathways has been shown, however, to be effectively blocked by prazosin, a selective α_1-adrenergic receptor antagonist.

In the cerebral cortex of adrenalectomized rats an increase in noradrenaline-stimulated cAMP accumulation has been reported by several groups (Mobley and Sulser, 1980; Mobley *et al.*, 1983; Duman *et al.*, 1989). The potentiation of the adenylyl cyclase pathway is primarily the result of glucocorticoid insufficiency. Hypophysectomy mimicked the effects of adrenalectomy, while medullectomy was without effect. For adrenalectomy and hypophysectomy, replacement therapy with corticosterone abolished the potentiation of the noradrenaline-stimulated cAMP accumulation. Addition of ACTH, in constrast, reversed the effects of hypophysectomy, but *not* those of adrenalectomy. Interestingly, when the β-adrenergic agonist isoproterenol was used to stimulate cAMP accumulation, the potentiation induced by adrenalectomy was not observed. These interesting observations suggest that the potentiation observed in adrenalectomy operates via some adrenergic receptor-mediated pathway other than the β-adrenergic receptor. Dissection of this system revealed a complex regulation of G-protein-mediated pathways controlling adenylyl cyclase in the cortex. The increase in adrenaline-stimulated cAMP accumulation does not appear to be the result of changes in any of the known components of the stimulatory adenylyl cyclase pathway, including the cAMP-dependent phosphodiesterase. Analysis of α_1-, α_2- and β-adrenergic receptors by radioligand binding revealed no significant changes in either the dissociation constants (K_d) or the maximum binding (B_{max}) of any of these adrenergic receptor subtypes (Mobley *et al.*, 1983). Fluoride-stimulated, G_s-mediated adenylyl cyclase activity was found to be equivalent in sham-operated and adrenalectomized rats (Mobley *et al.*, 1983). However, immunoblot analysis of G-protein subunits and RNA blot analysis of their respective mRNAs did reveal intriguing alterations upon adrenalectomy. Although seemingly a paradox, $G_s\alpha$ levels decreased and $G_i\alpha$ levels increased in the adrenalectomized rats. This process appears to be dependent on the regulation of the steady-state mRNA expression of these two G-protein α subunits (Saito *et al.*, 1989). In contrast, β1 and β2 levels were unaltered by adrenalectomy. How these alterations in the expression of α subunits of G-proteins are linked to the enhanced noradrenaline-stimulated adenylyl cyclase remains to be elucidated. Greater changes in α subunit expression, in more localized areas, may have profound biochemical and physiological significance on neurotransmission.

'Cross-talk' among G-protein-linked receptors may play an important role in glucocorticoid action. β-Adrenergic receptors, and possibly other G_s-coupled receptors, may be phosphorylated via protein kinase C (Johnson *et al.*, 1990; Bouvier *et al.*, 1991). The inhibitory G-protein, G_i, may also be phosphorylated via protein kinase C activation (see references in Johnson *et al.* (1990) and Chapter 4). Phosphorylation of some of the components of the hormone-sensitive adenylyl cyclase by α_1-adrenergic receptors may explain the potentiation by α-adrenergic receptor-mediated pathways. Activation of protein kinase C by phorbol esters decreases the sensitivity (EC_{50}) and increases the V_{max} for agonist-stimulated adenylyl cyclase by two distinctly different modes. The decrease in sensitivity for adrenaline-stimulated adenylyl cyclase appears to be primarily due to a direct phosphorylation of β-adrenergic receptors by protein kinase C. The β_2-adrenergic receptor displays a consensus phosphorylation site for protein kinase C within the third cytoplasmic loop, a region that is critical for receptor–G-protein coupling and activation (Hausdorff *et al.*, 1990; Johnson *et al.*, 1990). Phosphorylation of $G_i\alpha$ by protein kinase C has been implicated as a possible cause for the increase in the V_{max} of adrenaline-stimulated adenylyl cyclase. Phosphorylation of $G_i\alpha$ by protein kinase C may attenuate, directly or indirectly, the activity of members of this particular family of G-proteins (Bushfield *et al.*, 1990). Thus, attenuation of $G_i\alpha$ activity by phorbol esters may relieve a tonic inhibition of adenylyl cyclase and result in a potentiation of the adrenaline-stimulated adenylyl cyclase activity.

Glucocorticoid treatment also modulates the hormone-sensitive adenylyl cyclase system in rat cerebral cortex. ACTH therapy produces effects similar to those observed with glucocorticoid treatment in control, but not in adrenalectomized, animals. Thus, the primary modulator of transmembrane signalling again appears to be the glucocorticoid output of the hypothalamic–pituitary–adrenal axis. Responses to chronic treatment with either ACTH or with glucocorticoids can be divided into two distinct phases, 'early' and 'late' (Duman *et al.*, 1987). The early phase (< 5 days) of glucocorticoid action in both control and in adrenalectomized rats appears to involve the α-adrenergic receptor pathway(s) which potentiate β-adrenergic receptor- and 2-chloroadenosine- (which also stimulates adenylyl cyclase via adenosine receptors in rat cortex) mediated responses. In response to ACTH or glucocorticoid therapy the contribution by the α-adrenergic receptor component declines by nearly half. In the late phase (> 10 days) of chronic glucocorticoid or ACTH administration there is an increase in the β-adrenergic receptor-stimulated cAMP accumulation, an effect that appears to be generalized for G_s-mediated pathways (Duman *et al.*, 1987). Vasoactive intestinal peptide (VIP), 2-chloroadenosine and forskolin-stimulated cAMP accumulation also increase during this phase of treatment. Interestingly, analysis of $G_s\alpha$ by immunoblotting and by cholera toxin-catalysed ADP-ribosylation as well as RNA blot analysis of $G_s\alpha$ mRNA revealed increases in both parameters following glucocorticoid treatment (Saito *et al.*, 1989). The opposite is true for $G_i\alpha$, $G_i\alpha2$ levels were found to decline in response to glucocorticoid treatment (Saito *et al.*, 1989), while the levels of β-subunits were unaltered.

Other regions of the brain, including the hippocampus and limbic system, are important targets for glucocorticoids. Although much of the biochemical basis for changes in responsiveness to neurotransmitters in these two areas is unknown, dramatic changes in G-protein-mediated pathways have been shown to result from adrenalectomy and glucocorticoid treatment (Harrelson *et al.*, 1987; Harrelson and McEwen, 1987). VIP is a potent stimulator of cAMP accumulation in many areas of the brain (Harrelson *et al.*, 1987). VIP-stimulated cAMP accumulation in several areas of the limbic system, including the hippocampus, amygdala and septum, is greatly enhanced in both adrenalectomized and hypophysectomized rats (Harrelson *et al.*, 1987; Harrelson and McEwen, 1987). In contrast, VIP-stimulated cAMP accumulation in frontal cortex and olfactory bulb is not significantly altered in adrenalectomized rats. In the hippocampus, the responsiveness to histamine was decreased while that to β-adrenergic agonists was increased by adrenalectomy (Harrelson and McEwen, 1987). The expression of serotonin receptors and muscarinic receptors is increased, while α_1-adrenergic receptor is unaltered in the hippocampus of adrenalectomized rats (Biegon *et al.*, 1985). These effects on neurotransmitter systems are reversed by treatment of adrenalectomized rats with glucocorticoids (Harrelson *et al.*, 1987). Forskolin-stimulated cAMP accumulation was unaltered by these changes in glucocorticoid levels, suggesting little change in the catalyst itself.

4 HEART

Very little is known of the effects of glucocorticoids on G-protein-mediated pathways in the heart. Few reports have analysed the components distal to the level of receptor. However, glucocorticoids clearly increase both the inotropic and left ventricular work index (Baxter and Forsham, 1972; Kauman, 1972). Adrenalectomy may alter β-adrenergic receptor number and/or sensitivity to catecholamines. Abrass and Scarpace (1981) reported that β-adrenergic receptor number increased by 60% within 6 h following adrenalectomy in myocardial membranes prepared from adrenalectomized female rats, as compared to their sham-operated controls. The increase in β-adrenergic receptor expression was sustained for up to seven days and could be reversed by cortisol administration. Interestingly, the increase in β-adrenergic receptor number is paradoxical, i.e. responsiveness to catecholamine stimulation is reduced, not enhanced, in hearts of adrenalectomized rats. Using a different time period (10–21 days postoperative) and male rats (compared to female rats used in the studies by Abrass and Scarpace), Davies *et al.* (1981) observed no changes in β-adrenergic receptor number or affinity. The EC_{50} for isoproterenol-stimulated adenylyl cyclase was increased slightly from $0.5\,\mu M$ to $1.5\,\mu M$ in adrenalectomized rats. However, the maximal stimulation of the adenylyl cyclase by isoproterenol or by sodium fluoride was unaltered (Davies *et al.*, 1981).

5 LIVER

Hepatic glucose production from glycogenolysis and gluconeogenesis are stimulated by both glucagon and adrenaline. Glucagon acts via a specific, G_s-coupled receptor in liver. Adrenaline, on the other hand, activates both hepatic α_1- and β_2-adrenergic receptors, stimulating Ca^{2+} mobilization and cAMP accumulation, respectively. Like that observed in cerebral cortex, both β_2- and α_1-adrenergic receptors contribute to catecholamine-stimulated functions. Activation of protein kinase A by cAMP potentiates the α_1-adrenergic receptor-mediated pathway, stimulating the mobilization of IP_3 and intracellular Ca^{2+} (Burgess *et al.*, 1991). In the fetal and neonatal rat, the predominant pathway mediating adrenergic control of glucose metabolism is a β_2-adrenergic receptor subtype. At the end of the third postnatal week, however, the α_1-adrenergic receptor pathway dominates glycogenolysis (Huff *et al.*, 1991). The developmental changes in responsiveness probably reflect two factors. First, developmental changes in sympathetic innervation of the liver result in a switch from adrenaline to noradrenaline. Second, maturation of the hypothalamic–pituitary–adrenocortical axis at the third week after birth results in a functional regulation of steroid secretion (Huff *et al.*, 1991). There are, however, differences observed between the sexes in the relative contribution that each adrenergic pathway contributes to the overall control of glucose metabolism. In male adult rats, α_1-adrenergic receptors play the predominant role in the stimulation of gluconeogenesis and glycogenolysis. In female rats, however, both α_1- and β_2-adrenergic receptors participate. For either glucagon- or adrenaline-stimulated glycogenolysis and gluconeogenesis, the activities of several rate-limiting enzymes are altered simultaneously through protein phosphorylation. Glycogenolysis is stimulated by activation of glycogen phosphorylase while glycogen synthesis is attenuated by the inactivation of glycogen synthase. Glucagon and adrenaline promote both of these effects via phosphorylation of the two enzymes.

Adrenalectomy markedly impairs α_1-adrenergic-, β_2-adrenergic- and glucagon-stimulated hepatic gluconeogenesis and glycogenolysis (Studer and Borle, 1984; Chan *et al.*, 1979a,b; Exton *et al.*, 1972, 1976). The adrenergic receptor pathway mediating hepatic glucose metabolism recovers, in part, due to an increased β_2 responsiveness in adrenalectomized animals (Huff *et al.*, 1991). In hepatocytes from adrenalectomized animals, Ca^{2+}, exchangeable Ca^{2+} and basal cAMP levels are all significantly higher than in sham-operated animals. As a consequence, the adrenaline-stimulated rise in Ca^{2+} and IP_3 accumulation is severely blunted in adrenalectomized animals (Studer and Borle, 1984; Borle and Studer, 1990). This effect results from a greatly diminished α_1-adrenergic receptor responsiveness in livers of adrenalectomized animals. At the level of receptor expression, α_1-adrenergic receptors have been reported to be either decreased slightly (Goodhardt *et al.*, 1982; Borle and Studer, 1990) or unaltered (Guellaen *et al.*, 1978) by adrenalectomy. Interestingly, the ability of guanine nucleotides to displace competition curves (GTP shifts) for G-protein-linked receptors was found to be abolished by adrenalectomy. These data suggest that the major defect responsible for the diminished α_1-adrenergic receptor responsiveness is an uncoupling of α_1-adrenergic receptors from G-protein(s). In contrast to the loss of responsiveness of the α_1-adrenergic receptor pathway, the β_2-adrenergic-stimulated cAMP accumulation is enhanced in livers from adrenalectomized rats. This potentiation of cAMP accumulation does not compensate fully, however, for the dramatic loss of the α_1-adrenergic receptor pathway. In fact, the β_2-adrenergic receptor pathway is attenuated in adrenalectomized animals at a point distal to cAMP accumulation (Bitensky *et al.*, 1970; Exton *et al.*, 1972; Leray *et al.*, 1973). β_2-Adrenergic receptor levels, in fact, are increased in rat liver membranes following adrenalectomy (Wolfe *et al.*, 1976; Goodhardt *et al.*, 1982; Guellaen *et al.*, 1978). Enhancement of GTP-promoted, agonist-specific shifts of receptor affinity in concert with the increased expression of β_2-adrenergic receptor appear to provide a likely explanation for the increase in adrenaline-stimulated cAMP accumulation observed in adrenalectomized animals (Wolfe *et al.*, 1976; Goodhardt *et al.*, 1982; Guellaen *et al.*, 1978).

Much less is known about the status of either G-protein subunits or effectors in the liver of adrenalectomized animals. Garcia-Sainz *et al.* (1988) examined the steady-state levels of $G_s\alpha$ and $G_i\alpha$ in livers from adrenalectomized rats. Unexpectedly, the levels of $G_s\alpha$ (measured by cholera toxin-catalysed ADP-ribosylation) were found to be decreased by 40% in liver membranes from adrenalectomized rats. Functional reconstitution of cholate extracts from liver membranes with S49 cyc^- cell membranes (lacking $G_s\alpha$) confirmed the decrease in $G_s\alpha$ activity indicated by cholera toxin-catalysed ADP-ribosylation. $G_i\alpha$ levels measured by pertussis toxin-catalysed ADP-ribosylation were found to nearly double in the adrenalectomized state. Adenylyl cyclase activity, in contrast, was unaltered in liver membranes by adrenalectomy (Garcia-Sainz *et al.*, 1988).

6 LUNG

Bronchodilation in the lung following administration of adrenaline results largely from stimulation of β_2-adrenergic receptors that are coupled to adenylyl cyclase. From a clinical standpoint, glucocorticoid therapy has been proven also to be useful in the treatment of asthma (Svedmyr, 1990). Long-term therapy for asthma with β-adrenergic agonists is compromised by agonist-induced desensitization and downregulation of β-adrenergic receptors (Holgate *et al.*, 1977). The agonist-induced desensitization and downregulation is a characteristic common to most, if not all, G-protein-like receptors (Hausdorff *et al.*, 1990; Wang *et al.*, 1991). In fact, an imbalance between the β-adrenergic receptor and the α-adrenergic/muscarinic receptor pathways has been proposed as a cause of asthma (Insel and Wasserman, 1990). The result of chronic β-adrenergic receptor stimulation is a loss of effectiveness by β-agonists (adaptation) in relieving the bronchoconstriction of asthma (Scarpace and Abrass, 1982). Used in tandem with β-adrenergic receptor agonists, glucocorticoids display synergism with respect to promotion of bronchodilation (Holgate *et al.*, 1977; Salonen, 1985; Matilla and Salonen, 1984). Glucocorticoids appear to have two major effects in the lung. Glucocorticoids enhance β_2-adrenergic receptor responses in the lung and also display potent anti-inflammatory effects.

From studies both *in vitro* and *in vivo* it is clear that glucocorticoids increase β-adrenergic activity in the lung via increased expression of β_2-adrenergic receptors. Several studies have linked glucocorticoids to increased steady-state expression of lung β_2-adrenergic receptors (Salonen and Matilla, 1984; Mano *et al.*, 1979). Treatment of cultured lung fibroblasts with glucocorticoids also increases the apparent rate of synthesis of β_2-adrenergic receptors (Fraser and Venter, 1980). As would be expected from the other studies of adrenalectomized animals, the steady-state expression of β_2-adrenergic receptors declines in lungs of animals following adrenalectomy (Mano *et al.*, 1979; Scarpace *et al.*, 1982).

7 MOLECULAR MECHANISMS

Much work has been performed in exploring the molecular mechanisms responsible for glucocorticoid-promoted upregulation of β_2-adrenergic receptors. Many investigators have utilized cell culture systems to examine the basis for the glucocorticoid-promoted increase in expression of β_2-adrenergic receptors. When DDT_1 MF-2 hamster *vas deferens*, Chinese hamster ovary (CHO), or 3T3 F442A cell lines are treated with glucocorticoids, a rapid rise in β_2-adrenergic receptor mRNA levels is observed (Collins *et al.*, 1988; Hadcock and Malbon, 1988b; Hadcock *et al.*, 1989b; Malbon and Hadcock, 1988; Feve *et al.*, 1990). For DDT_1 MF-2 cells, there is an early (1–4 h) transient, three-fold increase in β_2-adrenergic receptor mRNA during glucocorticoid treatment (Collins *et al.*, 1988; Hadcock and Malbon, 1988b). Following the increase in receptor mRNA levels, receptor mRNA levels decline to a new steady-state level, approximately double that observed in untreated cells (Hadcock and Malbon, 1988a). An increase in β_2-adrenergic receptor expression follows the increase in receptor mRNA (Figure 2). The glucocorticoid-induced increase in β_2-adrenergic receptor mRNA was blocked by the inhibitor of transcription, actinomycin D (Hadcock and Malbon, 1988b). No changes in β_2-adrenergic receptor mRNA stability are observed in response to glucocorticoids (Hadcock and Malbon, 1988b). The primary mechanism by which glucocorticoids increase β_2-adrenergic receptor mRNA levels is a steroid-promoted increase in the receptor gene transcription (Collins *et al.*, 1988; Hadcock *et al.*, 1989b). Measured by nuclear run-off transcription assays, a five-fold increase in transcription of the β_2-adrenergic receptor gene was observed in DDT_1 MF-2 cells treated with dexamethasone (Collins *et al.*, 1988; Hadcock *et al.*, 1989b). Several consensus sequences for glucocorticoid responsive elements (GRE) have been identified in the β_2-adrenergic receptor gene (Kobilka *et al.*, 1987; Emorine *et al.*, 1987). A GRE located in the 5′ non-coding region of the β_2-adrenergic receptor appears to be obligatory for the glucocorticoid-promoted increase in receptor gene transcription (Malbon and Hadcock, 1988).

β-Adrenergic agonists also regulate β_2-adrenergic receptors at the level of mRNA. Following a β-agonist-promoted early (∼1–2 h) transient increase in receptor mRNA, β_2-adrenergic receptor mRNA levels decline to 50% of control levels by 16 h (Hadcock and Malbon, 1988a; Collins *et al.*, 1989; Hadcock *et al.*, 1989a,b). The decline in receptor mRNA levels contributes to the agonist-induced downregulation of β_2-adrenergic receptors. An agonist-promoted destabilization of β-adrenergic receptor mRNA is the primary basis for the loss of receptor mRNA (Hadcock *et al.*, 1989b). When DDT_1 MF-2 cells are treated with glucocorticoids before, during, or after β-adrenergic agonist treatment, the decline in receptor mRNA is attenuated by steroid treatment (Hadcock *et al.*, 1989b). The increased rate of transcription promoted by glucocorticoids is, therefore, sufficient to overcome the β-agonist-promoted destabilization of β-adrenergic receptor mRNA (Hadcock *et al.*, 1989b).

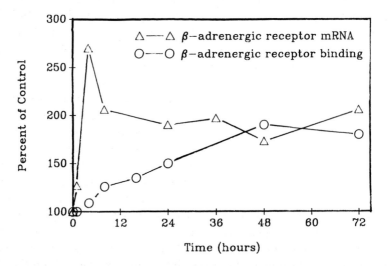

Figure 2. Dexamethasone increases the synthesis of β_2-adrenergic receptors: correlation between mRNA and protein levels. DDT_1 MF-2 cells were incubated with dexamethasone for the times indicated on the figure. At each time point β-adrenergic receptor levels were measured by equilibrium radioligand binding (open circles), or total cellular RNA was extracted and β_2-adrenergic receptor mRNA levels were measured by DNA excess solution hybridization assays (open triangles). Reprinted from Hadcock and Malbon (1988b).

Glucocorticoids appear to play a prominent role in the regulation of β_1- as well as β_2-adrenergic receptors. The glucocorticoid-induced changes in β_1-adrenergic receptor expression also appear to include regulation at the level of mRNA. In two cell lines used as models to study adipocyte differentiation and regulation, 3T3-L1 and 3T3 F442A, glucocorticoids decrease the steady-state expression of β_1-adrenergic receptors. These two cell lines express both β_1- and β_2-adrenergic receptors. Analysis of β_1-adrenergic receptor mRNA levels in 3T3-L1 fibroblasts revealed a 50% decrease in receptor mRNA levels when cells were treated with dexamethasone (Guest *et al.*, 1990). Treatment of 3T3-L1 cells with isobutyl-methyl xanthine and dexamethasone promotes differentiation to the adipocyte phenotype (Guest *et al.*, 1990). During differentiation the complement of β_1-adrenergic receptors and β_1-adrenergic receptor mRNA decline by 95%. Later studies suggested that glucocorticoids decrease β_1-adrenergic receptor mRNA levels and increase β_2-adrenergic receptor mRNA levels in 3T3-F442A cells differentiated to adipocytes by insulin (M10). Glucocorticoids also downregulate β_1-adrenergic receptors in C6 rat glioma cells, a process which also appears to be directed at the level of mRNA (Kiely *et al.*, submitted for publication). The mechanism by which glucocorticoids downregulate β_1-adrenergic receptor mRNA has not been established.

For most of the tissues examined, steady-state $G_s\alpha$ levels appear to be regulated both by adrenalectomy and glucocorticoid treatment. $G_s\alpha$ levels decline in adrenalectomy and are increased in response to glucocorticoid. This includes cortex, adipose tissue and liver. The functional significance of alterations in $G_s\alpha$ levels in some of these tissues remains to be elucidated. $G_s\alpha$ levels do not appear to be rate-limiting with respect to the output of the hormone-sensitive adenylyl cyclase system. For at least the β-adrenergic receptor-coupled adenylyl cyclase, receptor expression largely dictates the output, i.e. cAMP accumulation (George *et al.*, 1988). The level at which glucocorticoids regulate $G_s\alpha$ appears to include regulation at the level of mRNA. In the rat pituitary cell line GH_3, dexamethasone increases the steady-state $G_s\alpha$ expression $\sim$two-fold (Chang and Bourne, 1987). This increase in $G_s\alpha$ expression is preceded by a five-fold increase in $G_s\alpha$ mRNA. Inhibition of protein synthesis by cycloheximide prevented the glucocorticoid-induced increase in $G_s\alpha$ mRNA, suggesting that protein synthesis is necessary for the increase in mRNA levels (Chang and Bourne, 1987).

Limited information exists on the regulation of α subunits of other G-proteins by glucocorticoids. $G_i\alpha$ levels increase in frontal cortex of the brain and liver from adrenalectomized rats (Saito *et al.*, 1989;

Garcia-Sainz *et al.*, 1988). $G_i2\alpha$ mRNA levels increase in frontal cortex of adrenalectomized animals prior to the increase in $G_i\alpha$ expression (Saito *et al.*, 1989). In rat adipocytes, changes in mRNA levels also appear to be responsible for alterations in the expression of β subunits of G-proteins (Ros *et al.*, 1989b).

8 PERSPECTIVE

Studies of the physiology and pathophysiology of adrenal steroids, particularly the glucocorticoids, reveal profound and complex responses, largely tissue-specific in character. What happens in the brain, as compared to the liver or adipose tissue, can be quite distinct. Analysis of G-protein-mediated pathways has been made possible via the development of biochemical and molecular probes for the three classes of elements that comprise the transmembrane signalling devices, namely receptors, G-proteins, and effectors. In some cases glucocorticoid action can be best explained by changes at but one locus (i.e. α-adrenergic receptors in the frontal cortex), while in other cases the effects of the steroid appear to involve receptors, G-proteins, and perhaps effectors too (adipocytes). Clearly much work needs to be done, both *in vitro* and *in vivo*, on the regulation of transmembrane signalling by glucocorticoids, especially on the tissue-specific alterations in the components of G-protein-mediated pathways. Molecular cloning and analysis of gene structure provide insights into steroid regulation of transcription, identifying the existence of putative GREs within target genes. Post-transcriptional regulation (mRNA stability) and post-translational (protein turnover), too, are likely targets for steroid hormone regulation of G-protein-mediated pathways. Developing a complete knowledge of the physiological regulation of G-protein-mediated signalling by glucocorticoids clearly will require integration of information, not only the levels of molecular and cell biology, but also the physiology of the actions of these steroids.

ACKNOWLEDGEMENTS

This work was supported in part by United States Public Health Services Grants DK25410 and DK30111 from the National Institutes of Health and Grant-in-Aid 900663 from the American Heart Association.

C.C.M. is a recipient of Career Development Award K04 AM00786 from the NIH. J.R.H. is the recipient of a National Research Service Award T32 DK07521 from the NIDDK, NIH.

REFERENCES

Abrass, I.B. and Scarpace, P.J. (1981). *Endocrinology* **108**, 977–980.

Allen, D.O. and Beck, R.R. (1972). *Endocrinology* **91**, 504–510.

Bahouth, S.W. and Malbon, C.C. (1988). *Mol. Pharmacol.* **34**, 318–326.

Baxter, J.D. and Forsham, P.H. (1972). *Am. J. Med.* **53**, 573–577.

Biegon, A., Rainbow, T.C. and McEwen, B.S. (1985). *Brain Res.* **332**, 309–314.

Bitensky, M.W., Russell, V. and Blanco, M. (1970). *Endocrinology* **86**, 154–159.

Bojanic, D., Jansen, J.D., Nahorsky, S.R. and Zaagsma, J. (1985). *Br. J. Pharmacol.* **83**, 131–137.

Bokoch, G.M. (1987). *J. Biol. Chem.* **262**, 589–594.

Borle, A.B. and Studer, R.K. (1990). *J. Biol. Chem.* **265**, 19495–19501.

Bouvier, M., Guilbault, N. and Bonin, H. (1991). *FEBS Lett.* **279**, 243–248.

Braun, T. and Hechter, O. (1970). *Proc. Natl. Acad. Sci. USA* **66**, 995–1001.

Burgess, G.M., Bird, G.S.J., Obie, J.F. and Putney, J.W. Jr (1991). *J. Biol. Chem.* **266**, 4772–4781.

Bushfield, M., Murphy, G.J., Lavan, B.E., Parker, P.J., Hruby, V.J., Milligan, G. and Housley, M.D. (1990). *Biochem. J.* **268**, 449–457.

Chan, T., Blackmore, P., Steiner, K. and Exton, J. (1979a). *J. Biol. Chem.* **254**, 2428–2433.

Chan, T., Steiner, K. and Exton, J. (1979b). *J. Biol. Chem.* **254**, 11374–11378.

Chang, F. H. and Bourne, H. R. (1987). *Endocrinology* **121**, 1711–1715.

Collins, S., Caron, M.G. and Lefkowitz, R.J. (1988). *J. Biol. Chem.* **263**, 9067–9070.

Collins, S., Bouvier, M., Bolanowski, M.A., Caron, M.G. and Lefkowitz, R.J. (1989). *Proc. Natl. Acad. Sci. USA* **86**, 4853–4857.

Davies, A.O., De Lean, A. and Lefkowitz, R.J. (1981). *Endocrinology* **108**, 720–722.

Duman, R.S., Strada, S.J. and Enna, S.J. (1989). *Brain Res.* **477**, 166–171.

Emorine, L.J., Marulo, S., Delavier-Klutchko, C., Kaveri, S.V. Durieu-Trautman, O. and Strosberg, A.D. (1987). *Proc. Natl. Acad. Sci. USA* **84**, 6995–6999.

Emorine, L.J., Marullo, S., Briend-Sutren, M.-M., Patey, G., Tate, K., Delavier-Klutchko, C. and Strosberg, A.D. (1989). *Science* **245**, 1118–1121.

Exton, J.H., Friedman, N., Hee-Aik, E.W., Brineaux, J.D., Corbin, J.D. and Park, C.R. (1972). *J. Biol. Chem.* **247**, 1732–1739.

Exton, J., Miller, T., Harper, H. and Park, C. (1976). *Am. J. Physiol.* **230**, 163–170.

Feve, B., Emorine, L.J., Briend-Sutren, M.M., Lasnier, F. and Strosberg, A.D. (1990). *J. Biol. Chem.*

Fraser, C.M. and Venter, J.C. (1980). *Biochem. Biophys. Res. Commun.* **94**, 390–397.

Garcia-Sainz, J.A., Huerta-Bahena, M.E. and Malbon, C.C. (1988). *Am. J. Physiol. (Cell Physiol.)* **256**, C384–C389.

George, S.T., Berrios, M., Hadcock, J.R., Wang, H.-y. and Malbon, C.C. (1988). *Biochem. Biophys. Res. Commun.* **150**, 665–672.

Gilman, A.G. (1987). *Annu. Rev. Biochem.* **56**, 615–649.

Goodhardt, M., Ferry, N., Geynet, P. and Hanoune, J. (1982). *J. Biol. Chem.* **257**, 11577–11583.

Green, A., Johnson, J.L. and Milligan, G. (1990). *J. Biol. Chem.* **265**, 5206–5210.

Guellaen, G., Yates Aggerback, M., Vauquelin, G., Strosberg, A.D. and Hanoune, J. (1978). *J. Biol. Chem.* **253**, 1114–1120.

Guest, S.J., Hadcock, J.R., Watkins, D.C. and Malbon, C.C. (1990). *J. Biol. Chem.* **265**, 5370–5375.

Hadcock, J.R. and Malbon, C.C. (1988a). *Proc. Natl. Acad. Sci. USA* **85**, 5021–5025.

Hadcock, J.R. and Malbon, C.C. (1988b). *Proc. Natl. Acad. Sci. USA* **85**, 8415–8419.

Hadcock, J.R., Ros, M. and Malbon, C.C. (1989a). *J. Biol. Chem.* **264**, 13956–13961.

Hadcock, J.R., Wang, H.-y. and Malbon, C.C. (1989b). *J. Biol. Chem.* **264**, 19928–19933.

Harrelson, A.L. and McEwen, B.S. (1987). *J. Neurosci.* **7**, 2807–2810.

Harrelson, A.L., Rostene, W. and McEwen, B.S. (1987). *J. Neurochem.* **48**, 1648–1655.

Hausdorff, W.P., Caron, M.G. and Lefkowitz, R.L. (1990). *FASEB J.* **4**, 2881–2889.

Haynes, R.C. (1990). In: *The Pharmacological Basis of Therapeutics* (Gilman, A.G., Rall, T.W., Nies, A.S. and Taylor, P. eds), 8th edn Pergamon Press, New York, pp. 1431–1462.

Holgate, S.T., Balcwin, C.J. and Tattersfield, A. (1977). *Lancet* ii, 375–377.

Huff, R.A., Seidler, F.J. and Slotkin, T.A. (1991). *Life Sci.* **48**, 1059–1065.

Insel, P.A. and Wasserman, S.I. (1990). *FASEB J.* **4**, 2732–2736.

Johnson, J.A., Gokas, T.J. and Clark, R.B. (1986). *J. Cyclic Nucleotide Protein Phosphorylation Res.* **11**, 199–216.

Johnson, J.A., Clark, R.B., Friedman, J., Dixon, R.A.F. and Strader, C.D. (1990). *Mol. Pharmacol.* **38**, 289–293.

Kauman, A.J. (1972). *Naunyn-Schmiedeberg's Arch. Pharmacol.* **273**, 134–153.

Kobilka, B.K., Frielle, T., Dohlman, H.G., Bolanowski, M.A., Dixon, R.A.F., Keller, P., Caron, M.G. and Lefkowitz, R.J. (1987). *J. Biol. Chem.* **262**, 7321–7327.

Labrie, F., Pelletier, G., Borgeat, P., Drouin, J., Ferland, L. and Belanger, A. (1976). *Frontiers Neuroendocrinol.* **4**, 63–93.

Labrie, F., Veilleux, R., Lefevre, G., Coy, D.H., Sueiras-Diaz, J. and Schally, A.V. (1982). *Science* **216**, 1007–1008.

Lacasa, D., Agli, B. and Giudicelli, Y. (1986). *Biochem. Biophys. Res. Commun.* **153**, 489–497.

Lands, A.M., Arnold, A., McAuliff, J.P., Luduena, F.P. and Brown, T.G., Jr (1967). *Nature* **214**, 597–598.

Leray, F., Chambaut, A.M., Perrenoud, M.L. and Hanoune, J. (1973). *Eur. J. Biochem.* **38**, 185–192.

Limbird, L.E. (1988). *FASEB J.* **2**, 2686–2695.

Longabaugh, J.P., Didsbury, J., Spiegel, A. and Stiles, G.L. (1990). *Mol. Pharmacol.* **36**, 681–688.

Malbon, C.C. and Hadcock, J.R. (1988). *Biochem. Biophys. Res. Commun.* **154**, 676–681.

Malbon, C.C., Rapiejko, P.J. and Watkins, D.C. (1988). *Trends Pharmacol. Sci.* **9**, 33–36.

Mano, K., Akbarzadeh, A. and Townley, R.G. (1979). *Life Sci.* **25**, 1925–1930.

Matilla, M.J. and Salonen, R.O. (1984). *Br. J. Pharmacol.* **83**, 607–614.

McEwen, B.S., Dekloet, E.R. and Rostene, W. (1986). *Physiol. Rev.* **66**, 1121–1188.

Meyer, J.S. (1985). *Physiol. Rev.* **65**, 946–1010.

Mobley, P.L. and Sulser, F. (1980). *Nature* **286**, 608–609.

Mobley, P.L., Manier, D.H. and Sulser, F. (1983). *J. Pharmacol. Exp. Ther.* **226**, 71–77.

Moxham, C.P. and Malbon, C.C. (1985). *Biochemistry* **24**, 6072–6077.

Perkins, J.P. and Moore, M.M. (1973). *J. Pharmacol. Exp. Ther.* **185**, 371–378.

Rapiejko, P.J., Watkins, D.C., Ros, M. and Malbon, C.C. (1990). *Biochim. Biophys. Acta.* **1052**, 348–350.

Ros, M., Northup, J.K. and Malbon, C.C. (1989a). *Biochem. J.* **257**, 737–744.

Ros, M., Watkins, D.C., Rapiejko, P.J. and Malbon, C.C. (1989b). *Biochem. J.* **260**, 271–275.

Saito, N., Guitart, X., Hayward, M., Tallman, J.F., Duman, R.S. and Nestler, E.J. (1989). *Proc. Natl. Acad. Sci. USA* **86**, 3906–3910.

Salonen, R.O. (1985). *Acta Pharmacol. Toxicol. (Copenh.)* **57**, 147–153.

Salonen, R.O. and Mattila, M.J. (1984). *Acta Pharmacol. Toxicol. (Copenh.)* **55**, 425–428.

Scarpace, P.J. and Abrass, I.B. (1982). *J. Pharmacol. Exp. Ther.* **223**, 327–331.

Studer, R.K. and Borle, A.B. (1984). *Biochim. Biophys. Acta* **804**, 377–385.

Svedmyr, N. (1990). *Am. Rev. Respir. Dis.* **141**, S31–S38.

Thotakura, N.R., De Mazancourt, P. and Giudicelli, Y. (1982). *Biochim. Biophys. Acta* **717**, 32–40.

Wang, H.-y., Hadcock, J.R. and Malbon, C.C. (1991). *Receptor* **1**, 13–33.

Watkins, D.C., Northup, J.K. and Malbon, C.C. (1989). *J. Biol. Chem.* **264**, 4186–4194.

Wolfe, B.B., Harden, T.K. and Molinoff, P.B. (1976). *Proc. Natl. Acad. Sci. USA* **73**, 1343–1347.

CHAPTER SIX

G-proteins in Obesity

NICOLE BÉGIN-HEICK and NORMA McFARLANE-ANDERSON
Department of Biochemistry, University of Ottawa,
Ottawa, ON K1H 8M5, Canada

1 INTRODUCTION

G-protein-linked signalling pathways regulate diverse systems in response to stimuli such as hormones, neurotransmitters, light and
odours, among others (Gilman, 1987; Birnbaumer *et al.*, 1990). In
addition to the G-proteins that regulate adenylyl cyclase, ion channels
and the sensory pathways, several other types of cellular G-proteins
(e.g. p21^{ras} (Barbacid, 1987)), other low (e.g. ADP-ribosylation factors
(Kahn and Gilman, 1986); Rab proteins (Goud *et al.*, 1990) and high
(Udrisar and Rodbell, 1990) molecular weight proteins have been
reported and new ones are still being discovered. The present discussion will be restricted largely to the G-proteins that regulate adenylyl
cyclase, since little is known about other functions or types of G-
proteins in relation to obesity.

2 OBESITY

Obesity is a multifactorial syndrome characterized by the deposition
of excessive amounts of triglycerides at various body sites. It is a
significant cause of morbidity and/or mortality, especially through
linkage to cardiovascular disease and diabetes. In contrast to fat
accretion destined to fuel subsequent energy requirements or to obesity induced by dietary or chemical means, the fat depots accumulated
as a result of genetic obesity in animals appear to be conserved during
periods of need, such as fasting (Bégin-Heick, 1982).

G-Proteins
ISBN 0-12-497515-1

In the obese animal, cell systems or organs involved in the regulation of energy stores are the principal sites for the anomalous fat
storage. Thus, adipose tissue in various anatomical locations and the
liver are the organs primarily affected although most of the body
tissues are altered. In general adipocytes are larger and more numerous and lipid accumulation is increased.

Genetically determined obesity in animal models has been the
subject of extensive investigation. Obesity occurs in various rodents
(Table 1). The yellow, ob/ob, db/db and KK mouse and the fa/fa rat
have been the most thoroughly studied (Bray and York, 1971, 1979). In
addition, several other models have become available relatively more
recently. The degree of obesity in these animal models varies from
moderate to severe. Apart from the obesity, each syndrome has
multiple manifestations and, at least in the case of the ob/ob and db/db
mouse, these are affected by the genetic background on which the
mutant gene is carried (Coleman, 1982). This complexity in the expression of the disease has complicated the interpretation of research
results and has hampered the identification of the primary cause of
each obesity syndrome.

Table 1. Models of genetically determined rodent obesity and their mechanisms of inheritance

Phenotype	Genetic locus	Inheritance	Reference
Mouse-KK		polygenic	Konello *et al.* (1957)
yellow		s.d.	Danforth (1927)
adipose	ad	s.r.	Falconer and Isaacson (1959)
diabetes	db	s.r.	Hummel *et al.* (1966)
fat	fat	s.r.	Coleman (1982)
obese	ob	s.r.	Ingalls *et al.* (1950)
tubby	tub	s.r.	Coleman (1982)
Rat-corpulent	cp	s.r.	Koletsky (1975)
Zucker	fa	s.r.	Zucker and Zucker (1961)

s.d. = single dominant; s.r. = single recessive.

The obese syndrome in the ob/ob and db/db mouse and fa/fa rat is
associated with hyperinsulinaemia and resistance to insulin (Bray and
York, 1979). The ob/ob and db/db mouse are hyperglycaemic, whereas
the fa/fa rat is normoglycaemic or only slightly hyperglycaemic
(Zucker and Antoniades, 1972). All three genetic mutations are associ-

ated with adrenal hyperplasia and hypercorticism, both of which develop early in life (Dubuc, 1977; Coleman and Burkhart, 1977; Bray and York, 1979). Anomalies of thyroid hormone status have also been reported (Ohtake *et al.*, 1977; Young *et al.*, 1984; Kaplan *et al.*, 1985).

Resistance to the lipolytic action of catecholamines, principally in adipose tissue(s), is a characteristic of all three mutants. In the ob/ob mouse, the decreased catecholamine-stimulated lipolysis correlates with lower levels of cAMP production in adipocytes (Bégin-Heick and Heick, 1977). The fact that obesity appears to be characterized by failures in at least two pathways that are dependent on cAMP–hormone-stimulated lipolysis and adipocyte proliferation (Ailhaud, 1990) has stimulated research on the pathways regulating cAMP production and action in obesity models.

3 ADIPOSE TISSUE G-PROTEIN IN ANIMAL MODELS OF OBESITY

The adipocyte plays a major role in the maintenance of normal energy balance and much research has been done on the hormonal and biochemical control of lipogenesis and lipolysis. Much of the research on the regulation of adenylyl cyclase in obesity has been stimulated by the finding that adipose tissue of obese animals fails to mobilize fatty acids from stored triglycerides in response to appropriate stimuli. For this reason, much of the research on the regulation of adenylyl cyclase in obese animals has concentrated on the adipocyte. It is noteworthy that the rat adipocyte was used as a model system in the early studies of guanine nucleotide-mediated regulation of adenylyl cyclase (Rodbell, 1975), a choice that proved propitious as it is relatively easy to demonstrate stimulation and inhibition of the enzyme by guanine nucleotides and other ligands in this tissue. Cells isolated from the epididymal fat pad of rats were used extensively while other depot sites were studied much less. This choice of fat depot was made essentially for reasons of convenience, due to the discrete localization and anatomical structure of the rat tissue. It is important, however, to study other fat depots, as the hormonal control of metabolism may vary with the anatomical site of the fat tissue (Smith *et al.*, 1979).

The advent of established cell lines such as 3T3-L1, a subclone of 3T3 cells (Green and Kehinde, 1974), and HGFu and Ob17, two cell lines derived from fat pads of lean $(+/?)$ and ob/ob mice respectively

(Négrel *et al.*, 1978; Forest *et al.*, 1983), and the development of culture systems for preadipocytes differentiating into mature adipocytes in primary and secondary cultures (Van and Roncari, 1977, 1978), offer new means of studying the regulation of adipose tissue metabolism and the regulation of adenylyl cyclase in obesity syndromes. Adipocytes growing in primary and secondary cultures have been used recently to study preadipocyte replication and maturation in the cp rat (Shillabeer *et al.*, 1990) and it is a promising system for exploring the regulation of adipose cell replication and maturation as well as the regulation of G-proteins in genetic obesity.

The following discussion will deal mostly with the findings in three models of genetic obesity, the ob/ob mouse, the db/db mouse and the fa/fa rat. Wherever appropriate, data accumulated from human tissue will be incorporated. Findings relating to G-proteins in adipose tissue and liver will be highlighted. Information relating to other tissues is scant, but what is available relating to the brain, kidney, testis and pancreatic islets will be summarized.

4 THE IDENTIFICATION AND QUANTIFICATION OF G-PROTEINS IN TISSUES OF NON-OBESE ANIMALS

Initially, the α subunits of G-proteins were studied by virtue of their ability to be ADP-ribosylated by bacterial toxins (Ribeiro-Neto *et al.*, 1987). Specific antibodies are now available that allow the study of the various subunits (α, β and γ) and their isoforms, using immunoblotting techniques (Goldsmith *et al.*, 1987; Mumby *et al.*, 1988; Ros *et al.*, 1988). cDNAs corresponding to the various isoforms have been cloned, allowing a study of the genetic expression of the subunits (Bray *et al.*, 1986; Codina *et al.*, 1986; Ashley *et al.*, 1987; Fong *et al.*, 1987; Gao *et al.*, 1987; Jones and Reed, 1987; Blatt *et al.*, 1988).

4.1 Adipose tissue

Rat epididymal adipose tissue contains the short and long forms of $G_s\alpha$, three forms of $G_i\alpha$ (α1, α2 and α3), detectable by ADP-ribosylation and immunodetection, as well as both β35 and β36 subunits (Green and Johnson, 1989; Longabaugh *et al.*, 1989; Mitchell *et al.*, 1989; Ros, 1989a,b; Haraguchi and Rodbell, 1990). Rat adipocyte membranes seem to be devoid of $G_o\alpha$ (Hinsch *et al.*, 1988). In human adipocytes, $G_i\alpha$ and $G_s\alpha$ (the 42-kDa isoform being the more abundant species) subunits have been identified (Ohisalo and Milligan, 1989; Ohisalo *et al.*, 1989). There is evidence for $G_o\alpha$, albeit at very low concentrations, and the peptide may have a slightly lower molecular mass than the brain peptide (Homburger *et al.*, 1987; Rouot *et al.*, 1989).

4.2 Cell lines

A catecholamine-sensitive adenylyl cyclase (Rubin *et al.*, 1977) and the two isoforms of $G_s\alpha$ (Lai *et al.*, 1981) as well as $G_i\alpha$, $G_o\alpha$ and the β subunits (Watkins *et al.*, 1987), are present in 3T3-L1 preadipocytes. The status of these subunits during differentiation to adipocytes is not clear. Lai *et al.* (1981) found that ribosylation with cholera toxin (indicative of $G_s\alpha$) increases on differentiation and Gierschik *et al.* (1986), using ribosylation and immunoblotting techniques, found no change in the levels of $G_i\alpha2$, but a 50% decrease per cell for the $G_o\alpha$ and $\beta36$ subunits. On the other hand, Watkins *et al.* (1987, 1989) found that both α (except for $G_i\alpha2$ which decreases) and β subunits increase but at different rates, leading to a decrease in the overall ratio of β to α subunits as a result of differentiation.

Steady-state levels of the mRNA for $G_i\alpha2$, $G_o\alpha$, $G_s\alpha$ and β subunits decline upon differentiation. In the case of the $G_s\alpha$ and β mRNAs, the decrease is more modest than for the others, suggesting that the subunits are independently regulated. Taken together, these data show a lack of correlation between the patterns of mRNA and proteins as a result of differentiation, in the adipogenic cell (Watkins *et al.*, 1989).

4.3 Liver

The G-protein complement in rat liver membranes has been shown to consist of $G_s\alpha$, $G_i\alpha$ (α_i2, α_i3), $G_o\alpha$ and two β subunits (35 and 36) (Jones and Reed, 1987; Bushfield *et al.*, 1990b; Rapiejko *et al.*, 1990). The $\beta35$ and $\beta36$ subunits of human liver have also been described (Codina *et al.*, 1986).

4.4 Brain

Brain has the highest levels of $G_o\alpha$ of all tissues so far studied; in addition, it contains $G_s\alpha$, $G_i\alpha$ (α_i1, α_i2 and α_i3), as well as $\beta35$ and $\beta36$

(Neer *et al.*, 1984; Sternweis and Robishaw, 1984; Fong *et al.*, 1987; Jones and Reed, 1987; Homburger *et al.*, 1987).

5 G-PROTEIN LEVELS IN ADIPOCYTES OF OBESE ANIMALS

5.1 The ob/ob mouse

The notion that G-protein levels and/or function may be altered in obesity stems largely from observations in the ob/ob mouse. In the epididymal fat pad of the ob/ob mouse, the failure to respond to β-adrenergic stimuli occurs in the presence of normal lipolytic machinery and normal receptor complement. Consequently, since the catalytic unit of adenylyl cyclase responds normally to activation by NaF, a defect at the level of the G-protein signalling system was suggested (Bégin-Heick and Heick, 1977; Dehaye *et al.*, 1977, 1978). Initially, it was hypothesized that G_i might be more abundant or more responsive to its modulators; however, indirect evidence was provided showing that the impaired lipolysis in the ob/ob mouse was not due to increased levels or reactivity of G_i (Dehaye *et al.*, 1985). Functional $G_s\alpha$ (Bégin-Heick, 1985) and $G_i\alpha$ (Bégin-Heick, 1985; Greenberg *et al.*, 1987) are present in adipocyte membranes of both lean and ob/ob mice as assessed by their ability to undergo ribosylation by cholera and pertussis toxin, respectively. The observation that guanine nucleotides do not inhibit adenylyl cyclase in obese membranes as they do in the lean led to the conclusion that the amount of $G_i\alpha$ is abnormally low in the obese or, alternatively, that its functioning is impaired. Since intermediate levels of activation and inhibition were observed with tissue from mice heterozygous for the ob gene, the defect was suggested to be linked to the expression of that gene (Bégin-Heick, 1985) (see Section 10). Further experiments showed that, in fact, the levels of α subunits of G_s and G_i, as measured by ADP-ribosylation and/or immunodetection, were both lower in adipocyte membranes of obese than of lean mice. Moreover, the lower levels could be ascribed to specific isoforms of the α subunits (Figures 1 and 2). Immunodetection with appropriate antibodies did not reveal any differences in the levels of the β subunits (Bégin-Heick, 1990). The γ subunits were not studied.

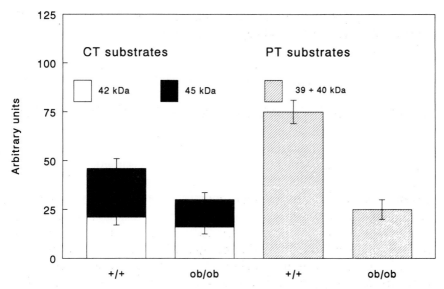

Figure 1. Quantification by ADP-ribosylation of the cholera and pertussis toxin substrates in adipocyte membranes of lean (+/+) and ob/ob mice.

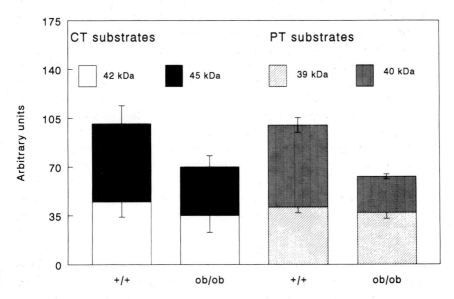

Figure 2. Quantification by immunodetection of the isoforms of $G_s\alpha$ and $G_i\alpha$ present in adipocyte membranes of lean (+/+) and ob/ob mice.

5.2 The db/db mouse

Adenylyl cyclase in adipocyte membranes from the db/db mouse is as resistant to activation by catecholamines as that in the ob/ob mouse. In contrast to ob/ob adipocytes, in the db/db there is evidence of an inhibitory effect of guanine nucleotides on adenylyl cyclase (Bégin-Heick and Coleman, 1988). ADP-ribosylation confirmed by immuno-detection showed that the various isoforms of the α subunits of G_s and G_i were present in similar proportions in adipocytes of the db/db mouse compared to its lean control (Bégin-Heick, submitted for publication). The resistance of adenylyl cyclase to catecholamines in adipose tissue of the db/db mouse can therefore not be ascribed to a lower abundance of the α subunits. No information is available on the β and γ subunits.

5.3 The fa/fa rat

In adipocytes from the obese Zucker rat, the rates of lipolysis and the response to hormone stimulation are much lower than in the lean rat (York and Bray, 1973). Vannucci *et al.* (1989) have shown that these changes are associated with increased adenosine receptor sensitivity to N^6-phenylisopropyl adenosine inhibition. The levels of $G_s\alpha$ as well as of all three isoforms of $G_i\alpha$ (α_i1, α_i2 and α_i3) estimated by immuno-blotting are lower than in the lean control but the ratio of $G_s\alpha$ to $G_i\alpha$ is higher in obese than in lean membrane preparations (Vannucci *et al.*, 1990). The findings of lowered $G_s\alpha$ and $G_i\alpha$ are consistent with those in the ob/ob mouse (Section 5.1). The fact that increased levels of GTP are required for stimulation by isoproterenol, coupled with the increased sensitivity of the A_i receptor in the fa/fa rat, suggests alterations in the function of the α subunits of G_s and G_i, but it is not yet known whether the defect is in the receptor or in the G-protein (Vannucci *et al.*, 1990).

5.4 Humans

There is an overall decrease in the amounts of both $G_s\alpha$ and $G_i\alpha$ subunits in membranes from markedly obese subjects compared to their lean controls (Ohisalo and Milligan, 1989; Richelsen, 1988), a situation comparable to that observed in the ob/ob mouse. Unlike the animal models, however, the lipolytic rates are higher in obese than in lean humans and the sensitivity to adenosine lower. Martin *et al.* (1990) found lowered adenylyl cyclase activity in adipocyte membranes from morbidly obese women. This was returned to normal in the post-obese state. Although G-protein levels were not assessed in that study, the authors inferred the alteration to be in the signalling system since stimulation by forskolin was normal. Because obesity is multifactorial and the obese phenotype may arise as a result of the modification of more than one gene, results in humans are even more difficult to interpret than those in inbred strains of animals.

5.5 Cell lines

Ob17 and HGFu are cell lines derived respectively from epididymal adipose tissue of obese (ob/ob) and lean (+/?) mice of the C57B1/6J strain (Négrel *et al.*, 1978). Little is known of the control of adenylyl cyclase and G-protein during differentiation and maturation of these cells. Data from immunochemical studies have shown that in the differentiated Ob17 cells there is less of the 45-kDa form of $G_s\alpha$ and more of the 42-kDa form than in the differentiated HGFu cell. These results are similar to those obtained with adipocyte membranes from ob/ob and lean mice (Section 5.1). The levels of expression of α_s, α_i2, α_o and β35 and β36 mRNA were the same in confluent HGFu and Ob17 cells. No differences were noted as a result of differentiation in the presence of insulin + triiodothyronine (Bégin-Heick *et al.*, 1991).

6 LIVER

The liver of obese rodents is enlarged and fatty, presumably due to increased lipogenesis which contributes to hypertriglyceridaemia and excessive lipid deposition. In comparison to the adipocyte, the magnitude of the inhibitory effects of guanine nucleotides and other ligands on liver adenylyl cyclase is small and appears to require the presence of monovalent cations (Jard *et al.*, 1981).

6.1 The ob/ob mouse

Paradoxically, the response of adenylyl cyclase to β-adrenergic catecholamines is greater in membranes from obese than from lean mice. Studies with α- and β-adrenergic agonists and antagonists indicate that the inhibitory effect of α-adrenergic agonists is not present in the membranes of obese animals, in spite of a normal complement of α-receptors measured by binding studies (Bégin-Heick and Welsh, 1988). Quantification by ADP-ribosylation revealed reduced amounts of both $G_s\alpha$ and $G_i\alpha$ in liver membranes from obese mice as compared to lean, as is the case in adipose tissue. Unlike adipose tissue, however, the levels of both the 45-kDa and 42-kDa forms of $G_s\alpha$ are lower in the obese liver. Two isoforms of $G_i\alpha$ (39 and 40 kDa) were detected by ADP-ribosylation; they appear to be present in similar amounts and were classified as α_i2 and α_i3 on the basis of their immunoreactivity. Probing with specific antibodies did not reveal detectable amounts of $G_i\alpha1$ or $G_o\alpha$ in liver membranes of lean or obese mice. Based on ADP-ribosylation results, the proportion of $G_s\alpha$ to $G_i\alpha$ is similar in liver from lean and ob/ob mice (1:10 and 1:12, respectively).

Studies at the mRNA level showed that the expression of the various isoforms is not different between lean and obese liver. They also demonstrate that the messages corresponding to $G_i\alpha1$ and $G_o\alpha$ are barely detectable, confirming the ADP-ribosylation and immuno-detection results. No differences were detected in the levels of the messages coding for the β subunits (McFarlane-Anderson *et al.*, 1992).

6.2 The db/db mouse

As was the case in adipocytes, there is no significant difference in the total levels of either $G_s\alpha$ or $G_i\alpha$ in liver membranes from the db/db mouse and its lean control (C57B1/Ks + / +), although the short form of $G_s\alpha$ was found in significantly greater amounts in the mutant. The isoforms of $G_i\alpha$ detected by immunological studies (α2 and α3) were present in approximately equal amounts (Bégin-Heick, unpublished data).

6.3 The fa/fa rat

In membrane preparations from livers of fa/fa rats, adenylyl cyclase activity as assessed with Gpp(NH)p in the presence of forskolin suggest that $G_i\alpha$ is either not present or not functional (Houslay *et al.*,

1989). ADP-ribosylation and immunoblotting revealed that the amounts of peptide are the same in lean and obese membranes, implying that the peptide is abnormal, perhaps as a result of post-translational modification. Bushfield *et al.* (1990a) recently identified $G_i\alpha2$ as the isoform involved and suggested that in the obese state, the protein is inactivated by phosphorylation via protein kinase C.

7 OTHER TISSUES

7.1 Brain

The complexity of the brain constitutes a major disadvantage in studies on the abundance and interaction of G-proteins. This is exacerbated in small rodents, particularly mice, where the lack of availability of sufficient tissue hampers a detailed study of specific brain structures. The interpretation of results is therefore subject to caution.

Membranes isolated from whole brain of lean or ob/ob mice contain similar amounts of both cholera- and pertussis-sensitive ADP-ribosylated peptides, as well as immunodetectable peptides. Whereas in liver and adipocyte membranes the two isoforms of $G_s\alpha$ are present in approximately equal amounts, in brain membranes, the long (45-kDa) isoform predominates. Furthermore, measurements of the levels of the message for each of the peptides support the finding that total brain G-proteins are unaltered in obesity (McFarlane-Anderson *et al.*, 1992). Because of the limitations noted above, it is possible that different conclusions would be reached, were specific structures studied. The best approach to this question might be the use of immunocytological techniques.

7.2 Testis

Obesity is often associated with gonadal hypofunction and/or infertility. In male ob/ob mice, there is active spermatogenesis in spite of reduced gonadal development (Bray and York, 1971). Mouse testis contains mRNA species corresponding to the short and long forms of $G_s\alpha$, to $G_i\alpha2$ and to the β subunits. The level of $G_s\alpha$ mRNA subunits is dramatically decreased in the ob/ob mouse compared to the lean. ADP-ribosylation and immunodetection show that while the levels of both isoforms of $G_s\alpha$ are low, that of the short (42-kDA) form is most severely affected (Figure 3). The lower levels in the obese mouse are not due to overall reduced protein synthesis, since the expression of the $G_i\alpha$ and β subunits is unchanged. Adenylyl cyclase activity in testicular membranes is low and there is no significant difference between membranes from lean and obese mice in the level of activity elicited by follicular stimulating hormone and/or guanine nucleotides (McFarlane-Anderson and Bégin-Heick, 1991). There is thus no apparent effect of the reduction in the $G_s\alpha$ subunits on cyclase activity in testicular tissue. It may be that G-protein subunits are present in large excess over the catalytic unit of adenylyl cyclase in these membranes so that even large changes in $G_s\alpha$ do not alter adenylyl cyclase kinetics (Pobiner *et al.*, 1985; Chang and Bourne, 1989).

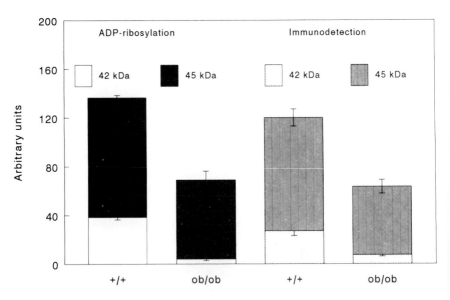

Figure 3. Quantification by ADP-ribosylation and immunodetection of the isoforms of $G_s\alpha$ present in testicular membranes of lean $(+/+)$ and ob/ob mice.

Testicular membranes from the db/db mouse did not show any changes in the $G_s\alpha$ subunits as assessed by ADP-ribosylation and immunodetection. The mRNA levels corresponding to each of the $G_s\alpha$, $G_i\alpha2$ subunits were also similar in the lean and db/db membranes (McFarlane-Anderson and Bégin-Heick, unpublished data).

7.3 Kidney

The levels of α and β subunits in kidney membranes were investigated in lean and ob/ob mice. Quantification by cholera and pertussis toxin ADP-ribosylation showed that the levels of both $G_s\alpha$ and $G_i\alpha$ subunits are unchanged in obesity (Figure 4). Immunodetection studies confirmed these findings. The expression of the various subunits, investigated by probing with appropriate cDNAs, is similar in both normal and mutant mice (McFarlane-Anderson and Bégin-Heick, 1991). No differences were found in kidney preparations of lean and db/db mice as assessed by ADP-ribosylation, immunodetection and cDNA probing (McFarlane-Anderson and Bégin-Heick, unpublished data).

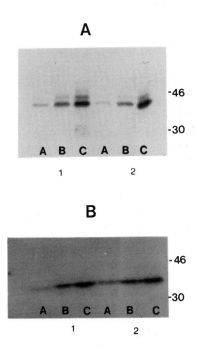

Figure 4. Incorporation of [^{32}P]-ADP-ribose by kidney membranes. In each case, 7.5 (A), 15 (B) and 30 (C) µg protein was subjected to ADP-ribosylation with cholera (a) or pertussis (b) toxin followed by electrophoresis and autoradiography. 1 = lean ($+/+$); 2 = obese (ob/ob). Size markers are in kDa.

7.4 Pancreatic islets

Many mutations leading to obesity are characterized by hyperinsulin-aemia due to an excessive secretory response of the β-cell of the islets of Langerhans to secretagogues. In particular, this has been docu-mented extensively in the ob/ob mouse. There is scant information on the identity and abundance of G-proteins in the insulin secretory cell. Limitations are imposed by the scarcity of the tissue and the fact that pancreatic islets are composed of several cell types of which the β-cell is, however, the most abundant. ADP-ribosylation experiments demon-strated that both $G_s\alpha$ and $G_i\alpha$ are present in the islet of the ob/ob as well as the lean mouse, but the studies were not done with the purpose of quantifying the subunits (Black et al., 1988). Indirect evidence, based on the kinetics of adenylyl cyclase and the effects of adrenaline and pertussis toxin on islet cAMP accumulation, suggests that the proportion and/or the function of G_i is altered in the ob/ob mouse. A major difficulty in interpreting data on insulin secretion is the uncer-tainty surrounding the nature and magnitude of the contribution of cAMP to the process. It is clear, for example, that the inhibitory effect of adrenaline on insulin secretion is independent from its effect on the adenylyl cyclase system (Ullrich and Wollheim, 1984; Black et al., 1988). Recently, it was documented that the inhibitory effect of somatostatin on insulin secretion occurs at least partially through a G-protein linked decrease in Ca^{2+} entry via voltage-dependent channels (Hsu et al., 1991). This may also be the mechanism for the inhibitory effect of adrenaline on insulin secretion.

8 FACTORS LIKELY TO AFFECT G-PROTEIN LEVELS IN OBESITY STATES

8.1 Hormonal effects

Thyroid and glucocorticoid hormones are considered permissive hor-mones because they alter tissue responses to catecholamines and insulin. They also exert influences on the expression of the α and β subunits of G-proteins, as well as on the β-adrenergic receptor sub-types in many cells (Malbon and Hadcock, 1988; Milligan et al., 1989; Ros et al., 1988, 1989a,b; Rapiejko et al., 1989; Guest et al., 1990, Levine et al., 1990a; Milligan and Saggerson, 1990).

8.1.1 Thyroid hormones

Due to defective deiodination of thyroxine (T_4) there is functional hypothyroidy in organs such as the brain, liver and brown adipose tissue in the ob/ob and db/db mouse and the fa/fa rat (Kaplan and Young, 1987; Kates and Himms-Hagen, 1990; Young *et al.*, 1984; Goldberg *et al.*, 1988; McIntosh *et al.*, 1989). The circulating levels of T_4 are reduced in the db/db but not in the ob/ob mouse, while the levels of triiodothyronine (T_3) are normal in both. In the fa/fa rat, tissue hypothyroidy is accompanied by low circulating levels of T_3 but normal levels of T_4. Impaired thyroxin 5'-deiodinase is responsible for the altered T_3/T_4 profiles in the three models. In fat cells from rats made hypothyroid by maintenance on iodine-deficient diets, the levels of peptides corresponding to $G_i\alpha$ and β, but not to $G_s\alpha$, increase (Ros *et al.*, 1988; Milligan and Saggerson, 1990; Levine *et al.*, 1990a). No change in the levels of $G_s\alpha$ or $G_i\alpha$ occurs in fat cells of thyroidectomized humans (Ohisalo and Milligan, 1989). More information is required before the importance of the thyroid status in the regulation of G-protein patterns in obesity is understood (see Chapter 7 for a full discussion).

8.1.2 Glucocorticoids

Since adrenalectomy reverses some of the features of obesity in db/db and ob/ob mice (Shimomura *et al.*, 1987) and in the fa/fa rat (Freedman *et al.*, 1986), corticosteroids are considered to be essential for the development of obesity. Glucocorticoids are reported to increase the expression of $G_s\alpha$ and adenylyl cyclase activity in GH_3 cells (Chang and Bourne, 1987). Consistent with this finding, Ros *et al.* (1989a,b) found that adrenalectomy decreases the steady-state levels of $G_s\alpha$ and β subunits in rat fat cells and that dexamethasone treatment reverses the effect of adrenalectomy. Both the ob/ob and the db/db mouse have adrenal hypertrophy and abnormally high circulating levels of corticosteroids (Bray and York, 1979), yet only the ob/ob mouse has altered G-protein patterns compared to its lean control (see Sections 5.1, 6.1 and 7.2). It is also noteworthy that in the ob/ob mouse the steady-state levels of $G_s\alpha$ are decreased (rather than increased, as would be predicted by the data of Ros *et al.* (1989a,b)) and its expression unchanged (in the liver) in spite of the prevalent hypercorticism. Conversely, in the adipose tissue of the db/db mouse, there is actually a significant increase of the 42-kDa isoform of $G_s\alpha$, although total levels of $G_s\alpha$ are not significantly different from the lean control. This latter finding is consistent with the notion that glucocorticoids have an enhancing effect on the levels of $G_s\alpha$, as reported by Ros et al. (1989a,b). Taken together, these findings indicate that the factor(s) responsible for the abnormally low levels of $G_s\alpha$ in the ob/ob mouse are distinct from the effect produced by glucocorticoids.

While adenylyl cyclase activity in adipocyte membranes of the ob/ob mouse is resistant to activation by catecholamines, it shows an enhanced response to ACTH compared to the lean. Adrenalectomy which corrects the hypercorticism and partially corrects the hyper-insulinaemia diminishes the response of adenylyl cyclase to ACTH but has no effect on catecholamine-stimulated activity in the obese. Conversely, in the lean, corticosteroid supplementation enhances adipocyte membrane response to ACTH (Bégin-Heick, 1987). It is therefore likely that the enhanced response to ACTH in the adipocyte of the ob/ob mouse is not due to altered G-protein function but rather to altered receptor function (see Chapter 5 for further information).

8.1.3 Insulin

The finding that insulin is able to inhibit cholera and pertussis toxin ribosylation in rat hepatocyte plasma membranes led to the suggestion that the insulin receptor interacts with G-proteins (Heyworth and Houslay, 1983; Heyworth et al., 1986; Rothenberg and Kahn, 1988). Some studies with rat adipocytes (Ciaraldi and Maisel, 1989; Davis and McDonald, 1990) support this hypothesis and further suggest that one subunit interacts with the insulin receptor. On the other hand, Joost et al. (1990), working with hepatocytes, found no evidence to support such an interaction. Streptozotocin diabetes is reported to lead to phosphorylation of the $G_i\alpha2$ subunit, rendering it nonfunctional (see Chapter 4). Bushfield et al. (1990a) postulated that such a loss of $G_i\alpha2$ may occur in the forms of diabetes found in the db/db mouse and the fa/fa rat. Recent evidence (Bushfield et al., 1990b) indicates that this may in fact be the case in the fa/fa rat. Streptozotocin administration results in insulinopoenia and the older (5–6 months) db/db mouse is insulinopoenic and less obese; in contrast, the fa/fa rat, the young (8–10 weeks) db/db mouse and the ob/ob mouse are hyperinsulinaemic. It is therefore not immediately evident how the same mechanism could explain the defect in all these models. In contrast to the (young) db/db mouse, there is a reduction in the amounts of $G_i\alpha$ and $G_s\alpha$ in liver membranes from the ob/ob mouse (Sections 6.1 and 6.2). The degree of phosphorylation of $G_i\alpha$ has not yet been assessed in either the db/db or the ob/ob mouse.

If there is interaction between the insulin receptor and one or more G-protein any perturbation in pattern and distribution of the subunits or receptors involved could be significant. It is of interest to note that translocation of insulin receptors to the Golgi apparatus occurs in the ob/ob mouse and Zucker rat (Kahn *et al.*, 1973; Posner *et al.*, 1978; Lopez *et al.*, 1988). Also, the receptor number is decreased in the liver of the db/db and ob/ob mouse in spite of increased levels of mRNA (Ludwig *et al.*, 1988).

It is evident that the effect of hormones on G-protein regulation is not simple. Furthermore, when dealing with whole animal systems as one does in studying obesity, it is difficult to separate the influence of one hormone from another, as there are intimate interrelationships between the various hormonal axes in the development and the maintenance of the obesity state. These problems cannot be resolved with studies in the whole animal or in tissue fractions. Primary or secondary culture systems where hormonal control can be studied in detail are required to understand the influences at play and the fine tuning involved in the hormonal regulation of the expression and/or steady state of the G-proteins in obesity.

8.2 Membrane localization

The mechanism of interaction between the G-proteins and the plasma membrane is not well understood (Buss *et al.*, 1987; Mumby *et al.*, 1990b). Recent evidence argues against an exclusive membrane localization of the various peptides (Wang *et al.*, 1989; Ercolani *et al.*, 1990; Brabet *et al.*, 1988). There is evidence for and against translocation of α subunits between membrane and cytoplasm as a result of stimulation (Iyengar *et al.*, 1988; Ransnas *et al.*, 1989; Rotrosen *et al.*, 1988; Premont and Iyengar, 1989; Haraguchi and Rodbell, 1990). In addition, Crouch (1991) and Crouch *et al.* (1990) report translocation of $G_i\alpha$ to the nucleus in cells induced to divide by growth factors.

The loss of $G_s\alpha$ subunits from the membrane after stimulation by cholera toxin (Chang and Bourne, 1989; Milligan *et al.*, 1989; Macleod and Milligan, 1989; Klinz and Costa, 1990), and of β subunit (Klinz and Costa, 1990) and $G_i\alpha$ subunits after prolonged exposure of rat adipocytes to the adenosine analogue N^6-phenylisopropyl adenosine (Green *et al.*, 1990) has been explained as accelerated degradation rather than translocation to the cytoplasm. It demonstrates, however, that the various G-proteins can be lost from the plasma membrane following stimulation. The importance of this phenomenon in the function of G-proteins as transducers is as yet uncertain.

The localization of the various subunits with respect to the membrane is dependent on hydrophobic interactions between the G-protein and the plasma membrane. Any alteration in the lipid microenvironment such as that reported for the adipocyte (York *et al.*, 1982) and for the liver and brain (Sena *et al.*, 1982) of ob/ob mice could therefore be significant in the membrane association and the functioning of G-proteins in obesity.

Immunocytochemical studies with HGFu and Ob17 (Bégin-Heick *et al.*, 1991) showed that subcellular localization of $G_i\alpha2$, $G_s\alpha$ and β was different: $G_i\alpha2$ co-localized with actin microfilaments at the plasma membrane, $G_s\alpha$ had a punctate cytoplasmic pattern and β had a pattern similar to $G_i\alpha2$ and $G_s\alpha$ and was also associated with the Golgi apparatus. Diverse localization of the G-protein subunits in the cell has also been noted in other systems (Brabet *et al.*, 1988; Gabrion *et al.*, 1989; Wang *et al.*, 1989; Toutant *et al.*, 1990; Holtzman *et al.*, 1991). This indicates that these proteins are not exclusively membrane bound, even in cells that have not been stimulated by ligands or agonists. No difference in localization was noted between lines derived from obese (OB17) and lean (GFu) mice.

8.3 Post-translational modifications

8.3.1 ADP-ribosylation

The discovery that bacterial toxins ADP-ribosylate G-protein α subunits was pivotal in the identification and subsequent purification and characterization of these peptides. Reports have now appeared suggesting that endogenous ADP-ribosylation of the α subunits occurs (Feldman *et al.*, 1987; Yamashita *et al.*, 1991). While the physiological significance of this observation is not known, it opens the possibility that such a mechanism may be involved in the cellular regulation of G-protein function. Although never demonstrated in the case of obesity, endogenous ADP-ribosylation could conceivably be a factor.

8.3.2 Phosphorylation

The phosphorylation of G-protein subunits by protein kinase C has been suggested to alter the availability of α subunits for normal function (Jakobs *et al.*, 1985; Katada *et al.*, 1985; Pyne *et al.*, 1989). The evidence available so far indicates a role for insulin as well as protein kinase C in this process (Davis and McDonald, 1990; Pyne *et al.*, 1989). Reports on the control of some G-protein subunits by phosphorylation in tissues of obese animals have already been mentioned (Section 8.1.3).

8.3.3 Myristoylation and isoprenylation

Myristoylation seems to be essential for the attachment of α subunits to the membrane, after their dissociation from the βγ subunits (Buss *et al.*, 1987; Jones and Spiegel, 1990; Mumby *et al.*, 1990b). Evidence has recently been presented that the α and γ subunits may be substrates for isoprenylation which may play a role in membrane association (Maltese, 1990; Maltese and Robishaw, 1990; Mumby *et al.*, 1990a). Defects in fat metabolism could conceivably alter this process and be a factor in obesity.

8.4 The γ subunit

The β and γ subunits exist as a tightly bound complex which can be separated only by detergent (Evans *et al.*, 1987; Hildebrandt *et al.*, 1985). Four forms of the γ subunit differing in their primary structure have been identified to date (Hurley *et al.*, 1984; Yatsunami *et al.*, 1985; Gautam *et al.*, 1989, 1990; Robishaw *et al.*, 1989). Additional heterogeneity in the γ subunits is introduced by post-translational modifications (Backlund *et al.*, 1990; Maltese and Robishaw, 1990; Mumby *et al.*, 1990a) resulting potentially in a wide array of forms for this particular subunit. It also appears that the γ subunits are much more tissue specific than the α or β subunits (Gautam *et al.*, 1990). It has been suggested by Gautam *et al.* that if structurally distinct βγ complexes had diverse functions, there could be a large increase in the variety of G-proteins. So far, there have been no reports on the identification and levels of γ subunits in tissues of obese animals. Recent reports that the γ subunits undergo post-translational modifications (Section 8.3) evoke the possibility that their insertion in the membranes of the obese animals could be affected by modifications of the lipid composition of the membrane due to obesity (Section 8.2).

9 GROWTH AND DIFFERENTIATION

Insulin (or insulin-like growth factor) is one of the elements considered essential for the differentiation of clonal 3T3-L1 cells into adipocytes (Hauner, 1990). The IGF-II receptor has been linked to $G_i\alpha2$ in Balb/c3T3 cells (Nishimoto et al., 1989) and there is evidence for a pertussis-sensitive component involved in the stimulation of DNA synthesis in other cell lines and hepatocytes (Chambard et al., 1987; Pouységur et al., 1988; Fujinaga et al., 1989; Spiegel, 1989; Crouch et al., 1990, Crouch, 1991). This places the G-proteins at crucial points in cell growth and differentiation, two processes that are essential for normal adipocyte development (Ailhaud, 1990).

Adipogenic cells are capable of undergoing differentiation/de-differentiation cycles in response to cytokines such as tumour necrosis factor (Torti et al., 1989), which act through G-proteins (Krönke et al., 1990). Alterations in the G-protein profile in obesity syndromes may thus contribute to the loss of control of cell accretion that is characteristic of many of these syndromes.

Table 2. Chromosomal localization of the G-protein subunits and the obese mutations

	Chromosome number		
	Human	Mouse	Rat
$G_i\alpha1$	7	n.m.	n.m.
$G_i\alpha2$	3	9	n.m.
$G_i\alpha3$	1	n.m.	n.m.
$G\alpha_s$	20	2	n.m.
β_1	1	19	n.m.
β_2	7	n.m.	n.m.
β_3		12	
db/db mouse		4	
ob/ob mouse		6	
ad/ad mouse		7	
tub/tub mouse		7	
fa/fa rat			n.m.

n.m. indicates that gene has not been mapped.

10 GENETIC ASPECTS

The obesity syndromes in the most frequently studied models involve single recessive mutations (Leiter, 1989). The known loci are indicated in Table 2. The genes coding for the various G-protein subunits (Blatt *et al.*, 1988; Levine *et al.*, 1990b) map to chromosomes that are different from either the db or ob genes. Products of the genes responsible for obesity or other factors present in the internal milieu could, however, influence the production and functional amounts of the G-protein subunits. The finding of a gene dosage effect of the ob gene on adenylyl cyclase activity supports this contention (Bégin-Heick, 1985).

Table 3. Levels of G-proteins in mouse and rat tissues assessed by ADP-ribosylation and/or immunodetection (protein) and Northern blotting (mRNA)

Animal	Tissue	Protein			mRNA		
		$G_s\alpha$	$G_i\alpha$	β	$G_s\alpha$	$G_i\alpha$	β
ob/ob mouse	Adipocyte[a]	↓	↓	→	nd	nd	nd
	Liver[b]	↓	↓	→	→	→	→
	Brain[b]	→	→	nd	→	→	→
	Testis[c]	↓	→	→	↓	→	→
	Kidney[c]	→	→	→	→	→	→
db/db mouse	Adipocyte[d]	→	→	nd	nd	nd	nd
	Liver[e]	→	→	nd	→	→	→
fa/fa rat	Adipocyte[f]	↓	↓	nr	nr	nr	nr
	Liver[g]	→	→	nr	nr	nr	nr

[a] Bégin-Heick (1990); [b] McFarlane-Anderson *et al.* (1992); [c] McFarlane-Anderson and Bégin-Heick (1991); [d] Bégin-Heick (submitted for publication); [e] Bégin-Heick and McFarlane-Anderson (unpublished data); [f] Vannucci *et al.* (1990); [g] Houslay *et al.* (1989), Bushfield *et al.* (1990); ↓ indicates that the isoform is lower than in the same tissue in the control animal; → indicates that the levels are the same in lean and obese tissue; nd, not done; nr, not reported.

So far, there is no pattern emerging that could allow us to elaborate a model of the effect of the obese state on G-protein pattern (Table 3), although there is a general agreement that, at least in some forms of obesity, there is an alteration of G-protein status and/or function. The lack of correlation which has been noted in several quarters (Watkins *et al.*, 1989; Lee *et al.*, 1989; Longabaugh *et al.*, 1989) between G-protein abundance measured at the protein level (by immunodetection and ADP-ribosylation) and genetic expression measured by the mRNA level, indicates that complex factors are at play. Except for the testis, mRNA levels corresponding to the isoforms of G-proteins were similar in tissues of lean and ob/ob mice, even when the peptides to which they correspond were found to be present in lower amounts in the ob/ob mouse. This may indicate that the efficiency of translation or the stability of mRNA is altered. Similarly, the rate of degradation of the peptides (Silbert *et al.*, 1990) and/or their association with the membrane (Section 8.2) could differ. The use of membrane preparations to quantify subunit abundance could also underestimate the actual amounts of the G-proteins if they are not exclusively membrane bound. Cross-regulation of the steady-state level of G-protein subunits has been invoked to explain the behaviour of other systems (Chang and Bourne, 1989; Hadcock *et al.*, 1990; Levine *et al.*, 1990a; Ransnas *et al.*, 1989; Haraguchi and Rodbell, 1990) and could also explain some of the alterations observed in obesity.

It will be of interest to determine whether any of the changes observed in obese mutants play a role in the development of obesity or are a result of the syndrome.

So far research has focused on the links between G-proteins and cAMP production via adenylyl cyclase stimulation. However, it could be that other systems that involve G-proteins, such as ion channel regulation, are of equal or greater importance in the development of the obese syndrome. They remain to be explored.

Among the many areas that are yet to be investigated, and of primary inportance, is the identification of the various obese genes and their respective product(s). It could be that product(s) of the gene modify G-proteins either translationally or by regulating protein degradation.

Rather than clonal cell lines which have often undergone chromosomal rearrangement, the use of primary cultures of normal diploid cells is more likely to yield meaningful results. Although these cultures usually consist of mixed cell populations, they more closely resemble the *in vivo* condition than do clonal cells, especially if the intent is to study a genetic defect.

REFERENCES

Ailhaud, G. (1990). *Current Opinion Cell Biol.* **2**, 1043–1049.

Ashley, P.L., Ellison, J., Sullivan, K.A., Bourne, H.R. and Cox, D.R. (1987). *J. Biol. Chem.* **262**, 15299–15301.

Backlund, P.S., Simonds, W.F. and Spiegel, A.M. (1990). *J. Biol. Chem.* **265**, 15572–15576.

Barbacid, M. (1987). *Annu. Rev. Biochem.* **56**, 779–827.

Bégin-Heick, N. (1982). *Rev. Can. Biol. Exp.* **41**, 83–90.

Bégin-Heick, N. (1985). *J. Biol. Chem.* **260**, 6187–6193.

Bégin-Heick, N. (1987). *Mol. Cell. Endocrinol.* **53**, 1–8.

Bégin-Heick, N. (1990). *Biochem. J.* **268**, 83–89.

Bégin-Heick, N. and Coleman, D.J. (1988). *Mol. Cell. Endocrinol.* **59**, 171–178.

Bégin-Heick, N. and Heick, H.M.C. (1977). *Can. J. Physiol. Pharmacol.* **55**, 1320–1329.

Bégin-Heick, N. and Welsh, J. (1988). *Mol. Cell. Endocrinol.* **59**, 187–194.

Bégin-Heick, N., Cadrin, M. and McFarlane-Anderson, N. (1991). *FASEB J.* **5**, A455.

Birnbaumer, L., Yatani, A., Vandongen, A.M.J., Graf, R., Codina, J., Okabe, K., Mattera, R. and Brown, A.M. (1990). *Br. J. Pharmacol.* **30**, 13S–22S.

Black, M., Heick, H.M.C. and Bégin-Heick, N. (1988). *Am. J. Physiol.* **255**, E833–E838.

Blatt, C., Eversole-Cire, P., Cohn, V.H., Zollman, S., Fournier, R.E.K., Mohandas, L.T., Nesbitt, M., Lugo, T., Jones, D.T., Reed, R.R., Weiner, L.P., Sparkes, R.S. and Simon, M.L. (1988). *Proc. Natl. Acad. Sci. USA* **85**, 7642–7646.

Brabet, P., Dumuis, A., Sebben, M., Pantaloni, C., Bockaert, J. and Homburger, V. (1988). *J. Neurosci.* **8**, 701–708.

Bray, G.A. and York, D.A. (1971). *Physiol. Rev.* **51**, 598–646.

Bray, G.A. and York, D.A. (1979). *Physiol. Rev.* **59**, 719–809.

Bray, P., Carter, A., Simons, C., Guo, V., Puckett, C., Kamholz, J., Spiegel, A. and Nirenberg, M. (1986). *Proc. Natl. Acad. Sci. USA* **83**, 8893–8897.

Bushfield, M., Griffiths, S.L., Murphy, G.J., Pyne, N.J., Knowler, J.T., Milligan, G., Parker, P.J., Mollner, S. and Houslay, M.D. (1990a). *Biochem. J.* **271**, 365–372.

Bushfield, N., Pyne, N.J. and Houslay, M.D. (1990b). *Eur. J. Biochem.* **192**, 537–542.

Buss, J.E., Mumby, S.M., Casey, P.J. and Gilman, A.G. (1987). *Proc. Natl. Acad. Sci. USA* **84**, 7493–7497.

Chambard, J.C., Paris, S., L'Allemain, G. and Pouységur, J. (1987). *Nature (Lond.)* **326**, 800–803.

Chang, F.-H. and Bourne, H.R. (1987). *Endocrinology* **121**, 1711–1715.

Chang, F.-H. and Bourne, H.R. (1989). *J. Biol. Chem.* **264**, 5332–5357.

Ciaraldi, T.P. and Maisel, A. (1989). *Biochem. J.* **264**, 389–396.

Codina, J., Stengel, D., Woo, S.L.C. and Birnbaumer, L. (1986). *FEBS Lett.* **207**, 187–192.

Coleman, D.L. (1982). *Diabetes* **31** (suppl. 1), 1–6.

Coleman, D.L. and Burkhart, D.L. (1977). *Diabetalogia* **13**, 25–26.

Crouch, M.F. (1991). *FASEB J.* **5**, 200–206.

Crouch, M.F., Belford, D.A., Milburn, P.J. and Hendry, I.A. (1990). *Biochem. Biophys. Res. Commun.* 167, 1369–1376.

Danforth, C.H. (1927). *J. Hered.* 18, 153–162.

Davis, H.W. and McDonald, J.M. (1990). *Biochem. Biophys. Res. Commun.* 171, 53–59.

Dehaye, J.-P., Winand, J. and Christophe, J. (1977). *Diabetologia* 13, 553–561.

Dehaye, J.-P., Winand, J. and Christophe, J. (1978). *Diabetologia* 15, 45–51.

Dehaye, J.-P., Hebbelinck, M., Winand, J. and Christophe, J. (1985). *Horm. Metab. Res.* 17, 333–336.

Dubuc, P. (1977). *Horm. Metab. Res.* 9, 95–97.

Ercolani, L., Stow, J.L., Boyle, J.F., Holtzman, E.J., Lin, H., Grove, J.R. and Ausiello, D.A. (1990). *Proc. Natl. Acad. Sci. USA* 87, 4635–4639.

Evans, T., Fawzi, A., Fraser, E.D., Brown, M.L. and Northrup, J.K. (1987). *J. Biol. Chem.* 262, 176–181.

Falconer, D.S. and Isaacson, J.H. (1959). *J. Hered.* 50, 290–292.

Feldman, A.M., Levine, M.A., Baughman, K.L. and Van Dop, C. (1987). *Biochem. Biophys. Res. Commun.* 142, 631–637.

Fong, H.K.W., Armatruda, T.T., Birren, B.W. and Simon, M.I. (1987). *Proc. Natl. Acad. Sci. USA* 84, 3792–3796.

Forest, C., Czerucka, D., Grimaldi, P., Vannier, C., Négrel, R. and Ailhaud, G. (1983). In: *The Adipocyte and Obesity: Cellular and Molecular Mechanisms* (Angel, A., Hollenberg, C.H. and Roncari, D.A.K., eds), pp. 53–64. Raven Press, New York.

Freedman, M.R., Horwitz, B.A. and Stern, J.S. (1986). *Am. J. Physiol.* 250, R595–R607.

Fujinaga, Y., Morozumi, N., Sato, K., Tokumitsu, Y., Fujinija, K., Kondo, Y., Ui, M. and Okajima, F. (1989). *FEBS Lett.* 245, 117–121.

Gabrion, J., Brabet, P., Nguyen Than Dao, B., Homburger, V., Dumuis, A., Sebben, M., Rouot, B. and Bockaert, J. (1989). *Cell. Signalling* 1, 107–123.

Gao, B., Mumby, S. and Gilman, A.G. (1987). *Proc. Natl. Acad. Sci. USA* 84, 6122–6125.

Gautam, N., Baetscher, M., Aebersold, R. and Simon, M.L. (1989). *Science* 244, 971–974.

Gautam, N., Northup, J., Tamir, H. and Simon, M.I. (1990). *Proc. Natl. Acad. Sci. USA* 87, 7973–7977.

Gierschik, P., Morrow, B., Milligan, G., Rubin, C. and Spiegel, A. (1986). *FEBS Lett.* 199, 103–106.

Gilman, A.G. (1987). *Annu. Rev. Biochem.* 56, 615–650.

Goldberg, J.R., Ehrmann, B. and Katzeff, H.L. (1988). *Endocrinology* 122, 689–693.

Goldsmith, P., Gierschik, P., Milligan, G., Unson, C.G., Vinitsky, R., Malech, H.L. and Spiegel, A.M. (1987). *J. Biol. Chem.* 262, 14683–14688.

Goud, B., Zahraoui, A., Tavitian, A. and Saraste, J. (1990). *Nature (Lond.)* 345, 553–556.

Green, A. and Johnson, J.L. (1989). *Biochem. J.* 258, 607–661.

Green, A., Johnson, J.L. and Milligan, G. (1990). *J. Biol. Chem.* 265, 5206–5210.

Green, H. and Kehinde, O. (1974). *Cell* **1**, 113–116.

Greenberg, A.S., Taylor, S.I. and Londos, C. (1987). *J. Biol. Chem.* **262**, 4564–4568.

Guest, S.J., Hadcock, J.R., Watkins, D.C. and Malbon, C.C. (1990). *J. Biol. Chem.* **265**, 5370–5375.

Hadcock, J.R., Ros, M., Watkins, D.C. and Malbon, C.C. (1990). *J. Biol. Chem.* **265**, 14784–14790.

Haraguchi, K. and Rodbell, M. (1990). *Proc. Natl. Acad. Sci. USA* **87**, 1208–1212.

Hauner, H. (1990). *Endocrinology* **127**, 865–872.

Heyworth, C.M. and Houslay, M.D. (1983). *Biochem. J.* **214**, 547–552.

Heyworth, C.M., Grey, A.-M., Wilson, S.R., Hanski, E. and Houslay, M.D. (1986). *Biochem. J.* **235**, 145–149.

Hildebrandt, J.D., Codina, J., Rosenthal, W., Birnbaumer, L. and Neer, E.J. (1985). *J. Biol. Chem.* **260**, 14867–14872.

Hinsch, K.D., Rosenthal, E., Spicher, K., Binder, T., Gausepohl, H., Frank, R., Schultz, G. and Joost, H.G. (1988). *FEBS. Lett.* **238**, 191–196.

Holtzman, E.J., Soper, B.W., Stow, J.L., Ausiello, D.A. and Ercolani, L. (1991). *J. Biol. Chem.* **266**, 1763–1771.

Homburger, V., Brabet, P., Audigier, Y., Pantaloni, C., Bockaert, J. and Rouot, B. (1987). *Mol. Pharmacol.* **31**, 313–319.

Houslay, M.D., Gawler, D.J., Milligan, G. and Wilson, A. (1989). *Cell. Signalling* **1**, 9–22.

Hsu, W.H., Xiang, H., Rajan, A.S., Kunze, D.L. and Boyd, A.E. (1991). *J. Biol. Chem.* **266**, 837–843.

Hummel, K., Dickie, M.M. and Coleman, D.L. (1966). *Science* **153**, 1127–1128.

Hurley, J.B., Fong, H.K.W., Teplow, D.B., Dreyer, W.J. and Simon, M.I. (1984). *Proc. Natl. Acad. Sci. USA* **81**, 6948–6952.

Ingalls, A.M., Dickie, M.M. and Snell, G.D. (1950). *J. Hered.* **41**, 317–318.

Iyengar, R., Rich, K.A., Herberg, J.T., Premont, R.T. and Codina, J. (1988). *J. Biol. Chem.* **263**, 15348–15353.

Jakobs, K.H., Bauer, S. and Watanabe, Y. (1985). *Eur. J. Biochem.* **151**, 425–430.

Jard, S., Cantau, B. and Jakobs, K.H. (1981). *J. Biol. Chem.* **256**, 2603–2606.

Jones, D.T. and Reed, R.R. (1987). *J. Biol. Chem.* **262**, 14241–14249.

Jones, T.L.Z. and Spiegel, A.M. (1990). *J. Biol. Chem.* **265**, 19389–19392.

Joost, H.G., Schmitz-Salue, C., Hinsch, K.D., Schultz, G. and Rosenthal, W. (1990). *Eur. J. Biochem.* **172**, 461–469.

Kahn, C.R., Neville, D.M. and Roth, J. (1973). *J. Biol. Chem.* **248**, 244–250.

Kahn, R.A. and Gilman, A.G. (1986). *J. Biol. Chem.* **261**, 7906–7911.

Kaplan, M.M. and Young, J.B. (1987). *Endocrinology* **120**, 886–893.

Kaplan, M.M., Young, J.B. and Shaw, E.A. (1985). *Endocrinology* **117**, 1858–1863.

Katada, T., Gilman, A.G., Watanabe, Y., Bauer, S. and Jakobs, K.H. (1985). *Eur. J. Biochem.* **151**, 431–437.

Kates, A.-L. and Himms-Hagen, J. (1990). *Am. J. Physiol.* **258**, E7–E15.

Klinz, F.-J. and Costa, T. (1990). *Eur. J. Biochem.* **188**, 567–576.

Koletsky, S. (1975). Am. J. Pathol. 80, 129–142.

Konello, K., Nozawa, K., Tomita, T. and Ezaki, K. (1957). Bull. Exp. Animal 6, 107–112.

Krönke, M., Schütze, S., Scheurich, P., Meichle, A., Hensel, H., Thoma, B., Kruppa, G. and Pfizenmaier, K. (1990). Cellular Signalling 2, 1–8.

Lai, E., Rosen, O.M. and Rubin, C.S. (1981). J. Biol. Chem. 256, 12866–12874.

Lee, R.T., Brock, T.A., Tolman, C., Bloch, K.D., Seidman, J.G. and Neer, E.J. (1989). FEBS Lett. 249, 139–142.

Leiter, E.H. (1989). FASEB J. 3, 2231–2241.

Levine, M.A., Feldman, A.M., Robishaw, J.D., Ladenson, P.W., Ahn, T.G., Moroney, J.F. and Smallwood, P.M. (1990a). J. Biol. Chem. 263, 3553–3560.

Levine, M.A., Modi, W.S. and O'Brien, S.J. (1990b). Genomics, 8, 380–386.

Longabaugh, J.P., Didsbury, J., Spiegel, A. and Stiles, G.L. (1989). Mol. Pharmacol. 36, 681–688.

López, S., Desbuquois, B., Postel-Vinay, M.C., Benelli, C. and Lavau, M. (1988). Diabetologia 31, 922–927.

Ludwig, S., Muller-Wieland, D., Goldstein, B.J. and Kahn, C.R. (1988). Endocrinology 123, 594–600.

Macleod, K.G. and Milligan, G. (1989). Cellular Signalling 2, 139–151.

Malbon, C.C. and Hadcock, J.H. (1988). Biochem. Biophys. Res. Commun. 154, 676–681.

Maltese, W.A. (1990). FASEB J. 4, 3319–3328.

Maltese, W.A. and Robishaw, J.D. (1990). J. Biol. Chem. 265, 18071–18074.

Martin, L.F., Klim, C.M., Vannucci, S.J., Dixon, L.B., Landis, J.R. and LaNoue, K.F. (1990). Surgery 108, 228–235.

McFarlane-Anderson, N. and Bégin-Heick, N. (1991). Cell. Signalling (in press).

McFarlane-Anderson, N., Bailly, J.E. and Bégin-Heick, N. (1992). Biochem J. (in press).

McIntosh, M.K., Berdanier, C.D. and Kates, A.-L. (1989). FASEB J. 3, 1734–1740.

Milligan, G and Saggerson, E.D. (1990). Biochem J. 270, 765–769.

Milligan, G., Unson, C.G. and Wakelam, M.J.O. (1989). Biochem. J. 262, 643–649.

Mitchell, F., Griffiths, S.L., Saggerson, E.D., Houslay, M.D., Knowler, J.T. and Milligan, G. (1989). Biochem. J. 262, 403–408.

Mumby, S.M., Pang, I.-H., Gilman, A.G. and Sternweis, P.C. (1988). J. Biol. Chem. 263, 2020–2026.

Mumby, S.M., Casey, P.J., Gilman, A.G., Gutowski, S. and Sternweiss, P.C. (1990a). Proc. Natl. Acad. Sci. USA 87, 5873–5877.

Mumby, S.M., Heukeroth, R.O., Gordon, J.I. and Gilman, A.G. (1990b). Proc. Natl. Acad. Sci. USA 87, 728–732.

Neer, E.J., Lok, J.M. and Wolf, L.G. (1984). J. Biol. Chem. 259, 14222–14229.

Négrel, R., Grimaldi, P. and Ailhaud, G. (1978). Proc. Natl. Acad. Sci. USA 75, 6054–6058.

Nishimoto, I., Hata, Y., Ogata, E. and Kojima, I. (1989). *Proc. Natl. Acad. Sci. USA* **84**, 14029–14038.

Ohisalo, J.J. and Milligan, G. (1989). *Biochem. J.* **260**, 843–847.

Ohisalo, J.J., Vikman, H.L., Ranta, S., Houslay, M.D. and Milligan, G. (1989). *Biochem. J.* **264**, 289–292.

Ohtake, M., Bray, G.A. and Azukiwawa, M. (1977). *Am. J. Physiol.* **233**, R110–R115.

Pobiner, B.E., Hewlett, E.L. and Garrison, J.C. (1985). *J. Biol. Chem.* **260**, 16200–16209.

Posner, B.I., Raquidan, D., Josefsberg, Z. and Bergeron, J.J.M. (1978). *Proc. Natl. Acad. Sci. USA* **75**, 3302–3306.

Pouységur, J., Chambard, J.C., L'Allemain, G., Magnaldo, I. and Seuwen, K. (1988). *Phil. Trans. R. Soc. Lond. B* **320**, 427–436.

Premont, R.T. and Iyengar, R. (1989). *Endocrinology* **125**, 1151–1160.

Pyne, N.J., Murphy, G.J., Milligan, G. and Houslay, M.D. (1989). *FEBS Lett.* **242**, 77–82.

Ransnas, L.A., Svodoba, P., Jasper, J.R. and Insel, P.A. (1989). *Proc. Natl. Acad. Sci. USA* **86**, 7900–7903.

Rapiejko, P.J., Watkins, D.C., Ros, M. and Malbon, C.C. (1989). *J. Biol. Chem.* **264**, 16183–16189.

Rapiejko, P.J., Watkins, D.C., Ros, M. and Malbon, C.C. (1990). *Biochim. Biophys. Acta* **1052**, 348–350.

Ribeiro-Neto, F., Mattera, R., Grenet, D., Sekura, R.D., Birnbaumer, L. and Field, J.B. (1987). *Mol. Endocrinol.* **1**, 472–481.

Richelsen, B. (1988). *Metabolism* **37**, 268–275.

Robishaw, J.D., Kalman, V.K., Moomaw, C.R. and Slaughter, C.A. (1989). *J. Biol. Chem.* **264**, 15758–15761.

Rodbell, M. (1975). *J. Biol. Chem.* **250**, 5826–5834.

Ros, M.M., Northrup, J.K. and Malbon, C.C. (1988). *J. Biol. Chem.* **263**, 4362–4368.

Ros, M., Northrup, J.K. and Malbon, C.C. (1989a). *Biochem. J.* **257**, 737–744.

Ros, M., Watkins, D.W., Rapiejko, P.J. and Malbon, C.C. (1989b). *Biochem. J.* **260**, 271–275.

Rothenberg, P.L. and Kahn, C.R. (1988). *J. Biol. Chem.* **263**, 15546–15552.

Rotrosen, D., Gallin, J.I., Spiegel, A.M. and Malech, H.L. (1988). *J. Biol. Chem.* **263**, 10958–10964.

Rouot, B., Carrette, J., Lafontan, M., Lan Tran, P., Fehrentz, J.A., Bockaert, J. and Toutant, M. (1989). *Biochem. J.* **260**, 307–310.

Rubin, C.S., Lai, E. and Rosen, O.M. (1977). *J. Biol. Chem.* **252**, 3554–3557.

Sena, A., Rebel, G., Bieth, R., Hubert, P. and Waksman, A. (1982). *Biochim. Biophys. Acta* **710**, 290–296.

Shillabeer, G., Forden, J.M., Russel, J.C. and Lau, D.C.W. (1990). *Am. J. Physiol.* **258**, E368–E376.

Shimomura, Y., Bray, G.A. and Lee, M. (1987). *Horm. Metab. Res.* **19**, 295–299.

Silbert, S., Michel, T., Lee, R. and Neer, E.J. (1990). *J. Biol. Chem.* **265**, 3102–3105.

Smith, U., Hammersten, P., Björntorp, P. and Kral, J.G. (1979). *Eur. J. Clin. Invest.* **9**, 327–332.

Spiegel, S. (1989). *J. Biol. Chem.* **264**, 6766–6772.

Sternweis, P.C. and Robishaw, J.D. (1984). *J. Biol. Chem.* **259**, 13806–13813.

Strassheim, D., Milligan, G. and Houslay, M.D. (1990). *Biochem. J.* **266**, 521–526.

Torti, F.M., Torti, S.V., Larrick, J.W. and Ringold, G.M. (1989). *J. Cell. Biol.* **108**, 1105–1113.

Toutant, M., Gabrion, J., Vandaele, S., Peraldi-Roux, S., Barhanin, J., Bockaert, J. and Rouot, B. (1990). *EMBO J.* **9**, 363–369.

Udrisar, D. and Rodbell, M. (1990). *Proc. Natl. Acad. Sci. USA* **87**, 6321–6325.

Ullrich, S. and Wollheim, C.B. (1984). *J. Biol. Chem.* **259**, 4111–4115.

Van, R.L.R. and Roncari, D.A.K. (1977). *Cell. Tissue Res.* **181**, 197–203.

Van, R.L.R. and Roncari, D.A.K. (1978). *Cell. Tissue Res.* **195**, 317–329.

Vannucci, S.J., Klim, C.M., Martin, L.F. and LaNoue, K.F. (1989). *Am. J. Physiol.* **257**, E871–E878.

Vannucci, S.J., Klim, C.M., LaNoue, K.F. and Martin, L.F. (1990). *Int. J. Obesity*, **14**, 125–134.

Wang, H.-Y., Berrios, M. and Malbon, C.C. (1989). *Biochem. J.* **263**, 519–532.

Watkins, D.C., Northrup, J.K. and Malbon, C.C. (1987). *J. Biol. Chem.* **262**, 10651–10657.

Watkins, D.C., Rapiejko, P.J., Ros, M., Wang, H.Y. and Malbon, C.C. (1989). *Biochem. Biophys. Res. Commun.* **165**, 929–934.

Yamashita, A., Sato, E., Yasuda, H., Kurokowa, T. and Ishibashi, S. (1991). *Biochim. Biophys. Acta* **1091**, 46–50.

Yatsunami, K., Pandya, B.V., Oprian, D.D. and Khorana, H.G. (1985). *Proc. Natl. Acad. Sci. USA* **82**, 1936–1940.

York, D.A. and Bray, G.A. (1973). *Horm. Metab. Res.* **5**, 355–360.

York, D.A., Hyslop, P.A. and French, R. (1982). *Biochem. Biophys. Res. Commun.* **106**, 1478–1483.

Young, R.A., Fang, S.-L., Prosky, J. and Braverman, L.E. (1984). *Life Sci.* **34**, 1783–1790.

Zucker, L.M. and Antoniades, H.A. (1972). *Endocrinology*, **103**, 1320–1330.

Zucker, L.M. and Zucker, T.F. (1961). *J. Hered.* **52**, 275–278.

CHAPTER SEVEN

Thyroid Disorders

DAVID SAGGERSON
Department of Biochemistry and Molecular Biology,
University College London, Gower Street, London,
WC1E 6BT, UK

1 INTRODUCTION

Thyroid disorders are not rare, with hypothyroidism arising frequently in areas with iodine deficiency. Autoimmune processes are thought to be the basis for most other cases of both hypo- and hyperthyroid disorders. Overt hypothyroidism has a world-wide prevalence of approximately 1%. If subclinical cases are included, the overall prevalence is probably in excess of 5% (Bilous and Tunbridge, 1988). Tunbridge (1978) has reported a prevalence of hyperthyroidism in British women in excess of 2%, which is 10 times greater than that found in men.

The experimental studies described here are based on rat models of the disorders. Although in some early work the hypothyroid state was established by thyroidectomy, most recent work has centred on animals given drinking water containing propylthiouracil (PTU), usually in conjunction with an iodine-poor diet. Most studies of hyperthyroidism have been made with rats injected with triiodothyronine (T_3), or, occasionally, thyroxine (T_4). Usually this has been a daily regime extending over 3–7 days. Obviously the extent of thyroid hormone abnormality produced in these studies has varied between different laboratories. Nevertheless a number of clear, reproducible trends have become apparent.

I have opted for a tissue presentation, setting, where possible, observed changes into the more general physiological framework. This survey may appear to overconcentrate on the adipocyte. That is simply because it is from this cell that we have so far gained the most information in this field of study.

G-Proteins
ISBN 0-12-497515-1

2 ADIPOSE TISSUE

2.1 Adipose tissue metabolism

These comments are almost entirely concerned with white adipose tissue. Although thyroid disorders cause extensive changes in the regulation of brown adipose tissue, the possible involvement of G-proteins in such changes has not yet been addressed.

White adipose tissue, which occurs in numerous regions of the mammalian body, plays an important role in metabolic fuel disposition and selection. In essence its role is the storage of long-chain fatty acids as triacylglycerols in times of energy surfeit and the mobilization of fatty acids out of this triacylglycerol store in times of anticipated or actual energy demand. The functional cell in this regard is the adipocyte. Because of their relative ease of isolation (Rodbell, 1964) and because of their high sensitivity to various hormones both *in vivo* and *in vitro*, their study has contributed significantly to understanding cell signalling in general as well as its modification by thyroid disorders.

In a very general sense adipose tissue metabolism can be summarized as 'esterification versus lipolysis' (reviews: Saggerson, 1985, 1988). A certain amount of tissue turnover of triacylglycerol stores probably continues whatever the physiological state. However, whether adipose tissue is in a state of net deposition or net mobilization of stores depends upon the summation of several hormonal inputs. The regulation of triacylglycerol synthesis and deposition is beyond the scope of this chapter since roles for G-proteins in their regulation are not yet clearly established. The central feature of lipolysis, the mobilization process, is the hormone-sensitive lipase which is rate-limiting for the process. Hormone-sensitive lipase is activated by cAMP-dependent protein kinase through phosphorylation at a single regulatory serine residue (Stralfors and Belfrage, 1983; Stralfors *et al.*, 1984). Dephosphorylation and inactivation is mainly through actions of phosphoprotein phosphatases 2A and 2C (Olsson and Belfrage, 1987).

It has long been recognized that adipocyte adenylyl cyclase is stimulated by several hormones through the G_s pathway. The most frequently considered stimulators of lipolysis are adrenaline and noradrenaline, acting at the β-adrenergic receptor. In addition corticotropin and glucagon acting at their own receptors also activate adipocyte adenylyl cyclase and increase lipolysis in several species

(Londos *et al.*, 1981; Saggerson, 1986). Effects of these lipolytic hormones are antagonized by insulin; probably both through enhancement of cAMP phosphodiesterase activity (Elks *et al.*, 1983; Beebe *et al.*, 1985; Degerman *et al.*, 1990) and through increased dephosphorylation of the hormone-sensitive lipase (Stralfors and Honnor, 1989). Of particular relevance to this chapter, it is now realized that other agents inhibit fat cell adenylyl cyclase and inhibit lipolysis through pathways involving pertussis toxin-sensitive G-proteins (Moreno *et al.*, 1983; Murayama and Ui, 1983; Olansky *et al.*, 1983). Inhibitory receptors for adenosine (A_1 subtype), E-series prostaglandins and nicotinate (Londos *et al.*, 1981) together with α_2-adrenergic receptors (Rebourcet *et al.*, 1988; Garcia-Sainz and Medina-Martinez, 1989) are found on adipocytes. In the case of adenosine and the prostaglandins these inputs are probably paracrine or 'local' effects, possibly providing a degree of tonic inhibition of adenylyl cyclase. The inhibitory effect of adenosine, in particular, has had important implications for the design of properly defined studies of adipocyte adenylyl cyclase and lipolysis. This is because incubated adipocytes release adenosine (Schwabe *et al.*, 1973), whose effects are generally seen at low concentrations (10^{-10} to 10^{-8} M) (Ebert and Schwabe, 1973; Fain, 1973; Fain and Wieser, 1975; Saggerson, 1986). In addition nucleotidase activities on membranes will generate adenosine from exogenous ATP (e.g. in assays of adenylyl cyclase). Therefore, to make sense of experiments using fat cells or their plasma membranes it has become regarded as obligatory to remove endogenous adenosine by adding adenosine deaminase (Schwabe *et al.*, 1975; Fernandez and Saggerson, 1978). If regulatory input through the adenosine receptor is then to be studied, adenosine deaminase-resistant analogues such as N^6-phenylisopropyladenosine (PIA) have to be used. In early studies the 'adenosine problem' was not realized. Another complication arises from the fact that methylxanthines such as theophylline are potent antagonists at adenosine receptors. In early studies this was also not realized and the effects of these agents were misinterpreted.

2.2 Hypothyroidism and impaired stimulation of lipolysis

It has long been established that the normal lipolytic response of adipose tissue to adrenaline and other lipolytic hormones is severely impaired in hypothyroid states. Goodman and Knobil (1959) first showed that the adrenaline-induced mobilization of non-esterified fatty acids *in vivo* in rhesus monkeys is dependent upon optimal thyroid function. This finding was followed by numerous studies using incubated adipose tissue pieces from hypothyroid rats. In most cases enhancement of release of non-esterified fatty acids or glycerol from the tissue in response to high concentrations of adrenaline was negligible in hypothyroidism (Debons and Schwartz, 1961; Deykin and Vaughan, 1963; Krishna *et al.*, 1968). In other instances a partial response compared with the euthyroid state was observed (Bray and Goodman, 1965; Fisher and Ball, 1967). Studies of adrenaline dose–response curves revealed that hypothyroidism decreased the maximum lipolytic response as well as lessening the sensitivity of the response (Krishna *et al.*, 1968; Ichikawa *et al.*, 1971). Using isolated rat adipocytes, Correze *et al.* (1974) found that hypothyroidism also led to a defect in stimulation of lipolysis by corticotropin and by glucagon. Going on to study adenylyl cyclase in adipocyte membranes, Correze *et al.* (1974) observed that although maximal responses of the cyclase to glucagon, corticotropin and adrenaline were respectively decreased by 60%, 30% and 20% after thyroidectomy, the maximally stimulated activity of the cyclase with fluoride was the same as in membranes from euthyroid rats. A similar finding was made by Armstrong *et al.* (1974) and by Malbon *et al.* (1978a). In addition Malbon and Gill (1979) showed that activation of adipocyte adenylyl cyclase by cholera toxin was normal in hypothyroidism and Malbon and Graziano (1983) found that this state did not alter the maximum accumulation of cAMP by fat cells in response to forskolin. These studies, together with the demonstration that the level of cAMP-dependent protein kinase and the lipolytic response to dibutyryl cAMP are normal in hypothyroidism (Correze *et al.*, 1974), lead to the conclusion that the enzymic machinery for cAMP generation and for phosphorylation of the hormone-sensitive lipase is not itself impaired by thyroid insufficiency.

Several studies developed the theme that an elevated level of cyclic nucleotide phosphodiesterase was the reason for the impairment in hormone-mediated elevation of cAMP within intact adipocytes from hypothyroid animals. This idea arose from the observation that methylxanthine phosphodiesterase inhibitors (e.g. theophylline) restored maximal responses to adrenaline or corticotropin to levels comparable with those seen in the euthyroid state (Correze *et al.*, 1974; Armstrong *et al.*, 1974; Goswami and Rosenberg, 1978). However, although an increased phosphodiesterase activity in hypothyroidism has been confirmed by direct measurement (Correze *et al.*, 1974; Armstrong *et al.*, 1974; Van Inwegen *et al.*, 1975), it is likely that most of the effect of the methylxanthines in overcoming the lesion in activation of lipolysis is attributable to their action as adenosine receptor antagonists (see Section 2.3).

Another approach to attempt to gain understanding of the refractoriness of adenylyl cyclase to hormonal stimulation in the hypothyroid state was the measurement of abundance and affinity of adipocyte β-adrenergic receptors using the antagonist [^{3}H]dihydroalprenolol. Goswami and Rosenberg (1978) found that PTU-induced hypothyroidism had no effect on the B_{max} or K_D for dihydroalprenolol binding to intact adipocytes. Likewise, Malbon *et al.* (1978a), using the same hypothyroidism model, found no significant change in binding of this β-adrenergic receptor antagonist to an adipocyte membrane fraction. More recently the same group (Ros *et al.*, 1988), using quantitative immunoblotting, again found that PTU treatment did not cause any significant change in the abundance of β-adrenergic receptors. By contrast, Giudicelli (1978) found that thyroidectomy led to a three-fold decrease in the B_{max} for [^{3}H]dihydroalprenolol binding to membranes with no change in the K_D. In summary, different models of hypothyroidism may have differing effects on the abundance of adipocyte β-adrenergic receptors but no changes in the affinity of receptor–antagonist binding have been found.

Malbon (1980a) studied the displacement of [^{3}H]dihydroalprenolol from adipocyte membranes by isoproterenol. In the euthyroid case addition of 50 μM Gpp(NH)p caused a rightwards shift in the displacement curve (the EC_{50} for isoproterenol was increased from 1 μM to 7 μM). With membranes from hypothyroid rats the EC_{50} for isoproterenol displacement of the antagonist was 7 μM both in the absence and presence of Gpp(NH)p. This suggested that hypothyroidism decreased the affinity of the fat cell β-adrenergic receptor for agonists. Malbon (1980a) also concluded that hypothyroidism resulted in loss of ability of guanine nucleotide to regulate agonist affinity at the receptor and suggested that the hypothyroid cells displayed a reduced coupling of signal transduction from the β-adrenergic receptors to adenylyl cyclase.

Attempts to quantitate G-protein α subunits involved in the stimulatory pathway were first made by measurement of [^{32}P]ADP-ribosylation of adipocyte membrane proteins by cholera toxin. Malbon and Gill (1979) showed that hypothyroidism slightly (17–51%, depending on the toxin concentration) increased the relative abundance of the 42-kDa cholera toxin substrate. More recently, Rapiejko *et al.* (1989) found that hypothyroidism caused no significant change in the abundance of adipocyte membrane 42-kDa and 46-kDa forms of $G_s\alpha$ subunits as quantitated both by cholera toxin-catalysed [^{32}P]ADP-ribosylation and by immunoblotting. Likewise, Milligan and Saggerson (1990) found by immunoblotting that hypothyroidism caused non-significant decreases of 12% and 14% in the respective abundances of 42-kDa and 45-kDa forms of $G_s\alpha$ subunits (see Figure 1). Rapiejko et al. (1989) also quantitated the steady-state level of $G_s\alpha$ mRNA in rat adipocytes. This was done both by Northern blotting and by DNA excess solution hybridization analysis. Neither method revealed any significant change in $G_s\alpha$ mRNA level in hypothyroidism.

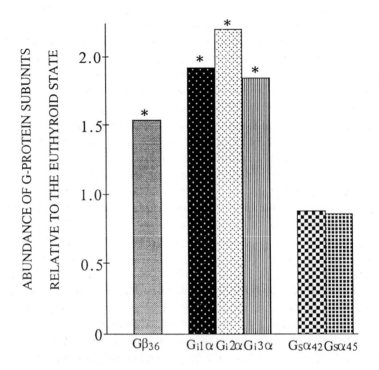

Figure 1. Effect of hypothyroidism on abundances of G-protein subunits in adipocyte plasma membranes. The data are taken from Milligan et al. (1987) and Milligan and Saggerson (1990). Adipocytes were isolated from the epididymal adipose tissues of euthyroid male rats and from similar animals made hypothyroid by administration of PTU and a low-iodine diet for four weeks. Plasma membranes were purified on Percoll gradients. G-proteins were quantitated after SDS-PAGE using anti-peptide sera. The values in the euthyroid state are set at 1.0. Statistically significant changes in hypothyroidism are indicated by *.

Malbon *et al.* (1984) conducted cell membrane hybridization studies to investigate the effect of hypothyroidism on G_s function. Lubrol extracts of adipocyte membranes from euthyroid and hypothyroid rats were reconstituted with membranes from S49 mouse lymphoma cyc⁻ mutant cells which lack functional G_s and therefore display little or no enhancement of adenylyl cyclase activity in response to GTP, fluoride or isoproterenol. Malbon *et al.* (1984) found that the G_s extracted from both normal and hypothyroid cells was equally effective in facilitating 'normal' response of the S49 cyc⁻ cyclase to stimulatory agents. The only possible indicator of some change in G_s to emerge from the study of Malbon *et al.* (1984) was that seen when adipocyte membranes were subjected to partial proteolysis with V8 protease after [³²P]ADP-ribosylation by cholera toxin. A small additional radiolabelled peptide unique to the hypothyroid state was consistently observed. The significance of this change is unclear and no differences in radiolabelled peptides were seen after partial proteolysis of euthyroid and hypothyroid membrane proteins by trypsin, chymotrypsin or elastase. At the present time there is no evidence to support conclusions other than that adipocyte G_s is essentially normal in the hypothyroid state, both in terms of abundance and function.

Malbon *et al.* (1984) performed three experiments designed to test the possibility that hypothyroidism affects the ability of the β-adrenergic receptor to interact productively with G_s. First, adipocyte membranes from hypothyroid rats were hybridized with *N*-ethylmaleimide-treated S49 cyc⁻ cell membranes possessing functional β-adrenergic receptors, but lacking functional G_s and adenylyl cyclase. A substantial activation of adenylyl cyclase by isoproterenol in this hybrid system was observed. However, when the cyc⁻ cell β-adrenergic receptor was inactivated by prior treatment with an affinity label, β-adrenergic responsiveness of the hybrid system cyclase was lost. Hence, although the cyc⁻ cell β-adrenergic receptor activated adipocyte G_s and adenylyl cyclase, the hypothyroid adipocyte β-adrenergic receptor did not. Second, β-adrenergic receptors from adipocyte membranes were hybridized with hepatocytes lacking functional β-adrenergic receptors. When the adipocyte β-adrenergic receptors had been derived from euthyroid animals, a cAMP response to isoproterenol was seen in the hybridized hepatocytes. By contrast, no significant response was seen if the donated β-adrenergic receptors had come from hypothyroid adipocytes. Third, pigeon erythrocyte [³²P]ADP-ribosylated G_s was reconstituted with rat adipocyte membranes which were then exposed to partial proteolysis by trypsin. During this proteolytic treatment isoproterenol enhanced the appearance of an ~30-kDa fragment, provided the β-adrenergic receptor bearing membranes were from euthyroid adipocytes. Since this effect was not observed with membranes from hypothyroid adipocytes, it was concluded that hypothyroidism impaired some agonist-promoted interaction between the adipocyte β-adrenergic receptor and G_s.

At this stage we have no further insight into the molecular basis of this coupling lesion, which is attributed to some selective modification of the adipocyte β-adrenergic receptor alone. The alternative possibility of some modification in G_s seems rather unlikely for reasons discussed above. Furthermore, although sensitivity of lipolysis to β-adrenergic stimulation is severely reduced in hypothyroidism, sensitivity of the process to corticotropin or glucagon is not. This is illustrated in Figure 2, which summarizes lipolysis regulation when inhibitory G_i-coupled input is abolished by addition of adenosine deaminase (this approach is discussed in Section 2.3). Under these conditions hypothyroidism caused a 40-fold increase in the EC_{50} for activation of lipolysis by noradrenaline but only four-fold increases in the EC_{50} values for corticotropin and glucagon.

In summary, hypothyroidism brings about extensive changes in the stimulatory regulation of adipocyte lipolysis. Changes in the various components are summarized in Table 1. With regard to G_s itself, there is no known intrinsic lesion but there is a failure of the β-adrenergic receptor to interact productively with this G-protein.

2.3 Hypothyroidism and enhanced inhibition of lipolysis

Through the 1970s studies of adipocyte adenylyl cyclase, cAMP accumulation and lipolysis were performed in the presence of unrealized and variable amounts of an endogenous G_i-coupled inhibitor of adenylyl cyclase, namely adenosine (see Section 2.1). Adipocytes and their membranes release or generate this potent inhibitory agonist in sufficient amounts to have severely complicated interpretation of earlier studies of this system and its perturbation by endocrine states. Figure 3 shows the consequences for adrenergic regulation of lipolysis on removal of endogenous adenosine by adenosine deaminase or the subsequent introduction of the agonist PIA.

Fernandez and Saggerson (1978), studying adrenalectomized rats, were the first to show that adenosine removal corrected an apparent lesion in the activation of lipolysis by noradrenaline or glucagon. It

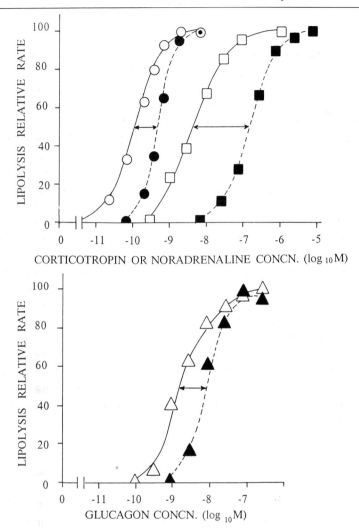

Figure 2. Dose–response curves for stimulation of lipolysis in adipocytes from euthyroid and hypothyroid rats. Animals were treated as described in Figure 1. Adipocytes were incubated with adenosine deaminase (1 unit/ml) in order to eliminate effects of endogenous adenosine. The data were taken from Saggerson (1986) and normalized. In this form the arrows indicate the shifts in the EC_{50} values between the euthyroid (open symbols) and the hypothyroid states (closed symbols). It is seen that hypothyroidism only modestly increased the EC_{50} values for corticotropin and glucagon by four-fold, whereas that for noradrenaline was increased by 40-fold. This is in accord with the conclusion of Malbon *et al.* (1984) that hypothyroidism impairs interaction between the β-adrenergic receptor and G_s. ○,●, corticotropin; □,■, noradrenaline: △,▲, glucagon.

was suggested that responsiveness to adenosine might be increased in this endocrine state (later directly confirmed by Saggerson (1980)). Following this, Ohisalo and Stouffer (1979) demonstrated that adenosine deaminase partially restored stimulation of lipolysis by adrenaline in adipocytes from hypothyroid rats. In particular, they found that the effect of PIA in inhibiting adrenaline-stimulated lipolysis was enhanced in hypothyroidism. This was manifest both as an increase in efficacy and as an approximate 10-fold decrease in EC_{50} for PIA. Figure 4 shows qualitatively similar findings by Saggerson (1986). Malbon and Graziano (1983) also concluded that adipocytes from hypothyroid rats are more sensitive to adenosine since adenosine deaminase overcame an apparent lesion in elevation of cAMP level by forskolin.

Table 1. Summary of changes in cell signalling components in rat adipocytes in hypothyroidism

Component	Comments
β-Adrenergic receptor abundance	(1) Decreased after thyroidectomy
	(2) Unchanged after PTU treatment
β-Adrenergic receptor performance	Impaired interaction with G_s
$G_s\alpha$ polypeptide abundance	Unchanged
$G_s\alpha$ mRNA abundance	Unchanged
$G_s\alpha$ functionality	Unchanged
Adenylyl cyclase activity	Unchanged
cAMP-dependent protein kinase activity	Unchanged
Cyclic nucleotide phosphodiesterase activity	Increased
Hormone-sensitive lipase activity	Decreased?
Adenosine receptor abundance	Slightly decreased
$G_i\alpha$ polypeptide abundance	Increased (G_i1, G_i2, G_i3)
$G_o\alpha$ polypeptide abundance	Increased
$G_i\alpha$ mRNA abundance	Increased (G_i2)
Gβ polypeptide abundance	Increased
Gβ mRNA abundance	Increased

Malbon *et al.* (1985) showed directly that increased sensitivity to PIA in hypothyroidism was seen at the level of its inhibition of cAMP accumulation in adipocytes in response to forskolin and at the level of inhibition of forskolin-stimulated adenylyl cyclase in fat cell membranes. Several lines of evidence suggested a post-receptor locus and implicated 'G_i' in this sensitivity change. First, increased sensitivity to

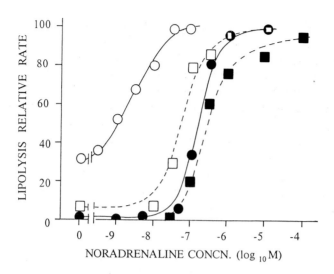

Figure 3. Dose–response curves for stimulation of lipolysis are substantially altered by low levels of adenosine. Data were taken from Fernandez and Saggerson (1978) and Saggerson (1986) and normalized. Adipocytes were obtained from euthyroid rats. ●, no additions (i.e. endogenous adenosine is present). ○, with adenosine deaminase (1 unit/ml). □, with adenosine deaminase (1 unit/ml) + 3×10^{-9} M PIA. ■, with adenosine deaminase (1 unit/ml) + 3×10^{-9} M PIA. The figure demonstrates: (a) Endogenous levels of adenosine are sufficient to cause a 100-fold rightwards shift in the EC_{50} for noradrenaline. (b) The situation with endogenous adenosine present is reproduced when the analogue PIA is added at levels in the range of 3–30 nM after removal of endogenous adenosine. (c) Basal lipolysis in the absence of noradrenaline is substantially increased when adenosine is absent. This effect is not seen in the hypothyroid state, suggesting an increased tonic inhibitory effect via 'G_i' (see Saggerson, 1986).

PIA was not reflected in a corresponding change in adenosine receptor status. Chohan *et al.* (1984) found that hypothyroidism increased neither the number nor the affinity of high- or low-affinity binding sites for [³H]PIA on purified adipocyte plasma membranes. Likewise, Malbon *et al.* (1985) found hypothyroidism to have no significant effect on B_{max} for binding of N^6-cyclohexyl-[³H]adenosine to membranes together with a slight decrease in affinity at a single class of sites. Second (Figure 4), hypothyroidism increased efficacy of and sensitivity to three separate G_i-coupled agonists that act through discrete receptors. Third, Malbon *et al.* (1985) found that GTP-induced inhibition of forskolin-stimulated adenylyl cyclase was also increased by hypothyroidism. Fourth, in hypothyroidism adipocyte lipolysis (Saggerson, 1986) and cAMP accumulation (Malbon *et al.*, 1985) shows resistance to the effect of pertussis toxin in attenuating the inhibitory effect of PIA. More indirectly, there is a substantial 'tonic' inhibition of basal lipolysis in hypothyroid fat cells even in the presence of adenosine deaminase (Saggerson, 1986). This is relieved by pertussis toxin, suggesting that it is caused by a change in 'G_i' (Saggerson, 1986).

Malbon *et al.* (1985) showed that pertussis toxin-catalysed [^{32}P]ADP-ribosylation of adipocyte membrane polypeptides in the 40–41-kDa region on SDS-PAGE was increased two- to three-fold in hypothyroidism and concluded that 'G_i' was increased in abundance in this state. This conclusion has been confirmed and extended through direct quantitation by immunoblotting. Milligan *et al.* (1987) used an anti-serum against the C-terminal decapeptide of rod transducin that recognized the α subunits of G_i1 and G_i2. Using electrophoretic conditions which could not resolve these 40-kDa and 41-kDa forms, they found an approximate two-fold increase in 'G_i' abundance relative to protein in Percoll-purified plasma membranes. Expressed per cell, the difference in 'G_i' abundance between the hypothyroid and euthyroid states was as much as five-fold. Ros *et al.* (1988) used antisera raised against oligomeric transducin or the α subunit of G_o (Rapiejko *et al.*, 1986) to probe immunoblots of adipocyte membranes. Hypothyroidism was found to increase the abundance of a 41-kDa band by approximately 50%. As discussed by Milligan and Saggerson (1990), this may correspond to $G_i1α$, to $G_i2α$ or to a combination of both. In addition, Ros *et al.* (1988) detected a 39-kDa band that was recognized by antisera to $G_oα$. The abundance of this was increased by ~70% in the PTU-induced hypothyroid model. By contrast, neither Mitchell *et al.* (1989), Rouot *et al.* (1989) nor Hinsch *et al.* (1988) have been able to detect substantial expression of authentic $G_oα$ in rat white adipocytes. With the realization by Mitchell *et al.* (1989) that rat adipose tissue contains mRNAs for the α subunits for G_i1, G_i2 and G_i3 and that adipocytes express all three α subunit forms in plasma membranes, Milligan and Saggerson (1990) used anti-peptide sera to show that hypothyroidism causes an approximate doubling of the abundance of all three $G_iα$ subunits in Percoll-purified rat adipocyte plasma membranes (Figure 1). Why adipocytes express all three $G_iα$ subunits is presently unknown. Further studies are necessary to establish which receptor/effector signalling pathways use which G_i. The significant change in abundance of all three α subunit types in hypothyroidism may have wider consequences beyond the regulation of lipolysis. This is because adenosine (presum-

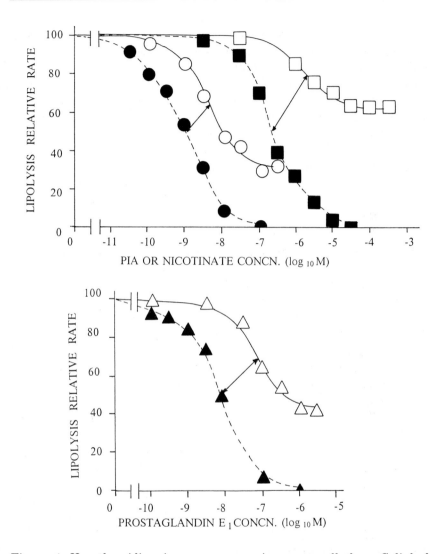

Figure 4. Hypothyroidism increases responsiveness to all three G_i-linked agonists PIA, prostaglandin E_1 and nicotinate. The data were taken from Saggerson (1986) and normalized. Animals were treated as described in Figure 1. Adipocytes were incubated with adenosine deaminase (1 unit/ml) and sufficient noradrenaline to stimulate lipolysis to 90% of maximal (0.05 µM and 1 µM noradrenaline in euthyroidism and hypothyroidism respectively). The figure shows that hypothyroidism increased the efficacy of all three agonists as well as decreasing the values for EC_{50} as indicated by the arrows. Open symbols, euthyroid; closed symbols, hypothyroid. ○,●, PIA; □,■, nicotinate: △,▲, prostaglandin E_1.

ably acting through G_i-coupled receptors) has potent effects on other adipose tissue processes as diverse as increasing the activities of lipoprotein lipase, pyruvate dehydrogenase and cyclic nucleotide phosphodiesterase as well as increasing glucose transport and amino acid metabolism (Ohisalo *et al.*, 1981; Joost and Steinfelder, 1982; Green, 1983; Honeyman *et al.*, 1983; Smith *et al.*, 1984; Wong *et al.*, 1985). Furthermore it has been noted that α-adrenergic generation of inositol phosphates in rat adipocytes is sensitive to pertussis toxin (Rapiejko *et al.*, 1986). The relevance of this signalling cascade to adipocyte metabolism is unclear at present, but it would be of interest to know if it also was altered in hypothyroidism.

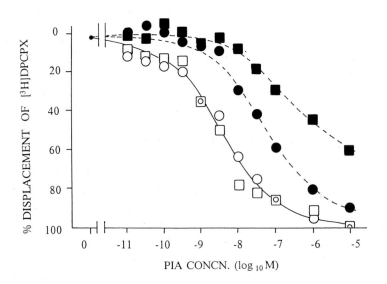

Figure 5. The effect of GTP to modify displacement of the antagonist [³H]DPCPX from adipocyte plasma membranes by the unlabelled agonist PIA is exaggerated in hypothyroidism. Animals were treated and adipocyte plasma membranes purified as described in Figure 1. Binding was measured by a filtration assay. 2 nM [³H]DPCPX and 1 mM dithiothreitol were present throughout together with the indicated concentrations of PIA without (open symbols) or with (closed symbols) 100 μM GTP. The figure shows that the EC_{50} for PIA (3 nM) was the same in the two states in the absence of GTP. By contrast, in the presence of GTP the EC_{50} was increased 29-fold to 105 nM in the euthyroid state and by 1100-fold to 3.2 μM in hypothyroidism. The data are from unpublished work (M.R. Orford and E.D. Saggerson). ○,●, euthyroid; □,■, hypothyroid.

A significant upregulation of adipocyte plasma membrane adenosine receptor-coupled G-protein function in hypothyroidism is also suggested by Figure 5. In the euthyroid state 100 µM GTP caused a 29-fold rightwards shift in the EC_{50} for displacement of the antagonist dipropylcyclopentylxanthine (DPCPX) by PIA. By contrast, in membranes from hypothyroid rats this GTP-dependent shift was increased to 1100-fold.

Accompanying the significant increase in $G_i\alpha$ subunits in the hypothyroid state in rats there is also a smaller increase in the abundance of G-protein β subunits. This was shown to be approximately 50% by Milligan *et al.* (1987) for the 36-kDa β subunit (Figure 1) and to be approximately 60% for 35–36-kDa β subunits by Ros *et al.* (1988). These changes in polypeptide abundance correlate well with a significant 40% increase in the steady-state level of mRNA for G-protein β subunits in adipocytes from hypothyroid rats (Rapiejko *et al.*, 1989). By contrast, Rapiejko *et al.* (1989) observed no significant change in the level of mRNA for $G_i2\alpha$ in hypothyroidism. It is therefore possible that the increased abundance of $G_i2\alpha$ polypeptide in hypothyroidism is due to altered translational or post-translational control, or is due to a decrease in the rate of turnover of the protein. Of possible relevance in this respect are the findings that treatment of rats (Parsons and Stiles, 1987), or primary culture of cells, with PIA (Green, 1987) decreases the abundance of the 41-kDa pertussis toxin substrate in adipocyte membranes with concomitant attenuation of the inhibition of forskolin-stimulated adenylyl cyclase by PIA or prostaglandin E_1. More recently Green *et al.* (1990) found that PIA causes downregulation of the α subunits of all three G_i subtypes in adipocyte primary cultures without altering the levels of the two forms of $G_s\alpha$ found in these cells. Could these effects relate to the upregulation of 'G_i' in the thyroid-deficient state? Supporting evidence is sparse at present but it is noteworthy that administration of T_3 or T_4 to hypothyroid rats significantly increased the adipose tissue content of adenosine (Ohisalo *et al.*, 1987). The implication was that the tissue adenosine level could be abnormally low in hypothyroidism, thereby lessening the heterologous desensitization process described by Parsons and Stiles (1987). In support of this conjecture, adipocyte 5'-nucleotidase ectoenzyme activity (AMP→adenosine) is decreased by 50% in hypothyroidism (Jamal and Saggerson, 1987). It should, however, be noted that the treatments with PIA that caused downregulation of 'G_i' (Parsons and Stiles, 1987; Green, 1987) also increased the activity of G_s and decreased the abundance of adenosine receptors in membranes. Since hypothyroidism does not decrease G_s activity (Malbon *et al.*, 1984) or increase adenosine receptor abundance (Chohan *et al.*, 1984) there is no reason to expect simple deprivation of extracellular adenosine to bring about changes identical to those in hypothyroidism.

As a precautionary note, the only study to date of adipocyte G-proteins in human hypothyroidism has not revealed any change in the membrane abundance of 'G_i' as recognized by an antiserum against the C-terminal decapeptide of rod transducin. This study (Ohisalo and Milligan, 1989) examined tissue from four patients that were thyroidectomized to remove carcinoma of the thyroid. At this stage it is not possible to tell whether upregulation of 'G_i' was not detected in this sample because of the genetic variability of the subjects and the euthyroid controls, because the measurements were made in a relatively crude $15\,000g$ membrane fraction (Mills *et al.*, 1986) or perhaps because rat and human 'G_i' respond differently to thyroid insufficiency.

In summary, upregulation of the G_i-coupled input to adipocytes contributes very substantially to the failure to activate cAMP elevation and lipolysis in hypothyroidism. At least in the PTU-treated rat model this may be explained by an increase in 'G_i' abundance. The mechanisms leading to this remain unknown.

2.4 Hyperthyroidism and altered regulation of lipolysis

Following the observation that plasma non-esterified fatty acid concentration was increased in human subjects with spontaneous or T_3-induced hyperthyroidism (Rich *et al.*, 1959), it was found in several laboratories that incubated fat pieces from rats treated with T_3 or T_4 showed an exaggerated lipolytic response to adrenaline or noradrenaline (Debons and Schwartz, 1961; Deykin and Vaughan, 1963; Bray and Goodman, 1965; Fisher and Ball, 1967; Krishna *et al.*, 1968). Where dose–response curves were obtained, both increased sensitivity and increased maximum response to the catecholamines were observed. The same trends were seen with isolated adipocytes (Caldwell and Fain, 1971; Mills *et al.*, 1986). When adipocyte cAMP elevation in response to catecholamines was investigated, an exaggerated response was also seen in hyperthyroidism (Caldwell and Fain, 1971; Malbon *et al.*, 1978a; Mills *et al.*, 1986). However, subsequent attempts to find a locus in the stimulatory pathway that is responsible for this upregulation have been unrewarding.

Quantitation of β-adrenergic receptors by binding of [³H]dihydro-alprenolol to adipocytes (Goswami and Rosenberg, 1978) or to adipocyte membranes (Malbon *et al.*, 1978a) revealed no changes in abundance in the hyperthyroid state. When [¹²⁵I]iodocyanopindolol binding was measured (Rapiejko and Malbon, 1987) there was a 26% increase in B_{max} but this was not matched by any significant change in the abundance of β-adrenergic receptors quantitated by immunoblotting (Ros *et al.*, 1988).

Rapiejko and Malbon (1987) studied the [³²P]ADP-ribosylation of adipocyte membrane proteins from T_3-treated rats by cholera toxin and found no change in the abundance of 42-kDa, 46-kDa or 48-kDa toxin substrates. The likelihood that $G_s\alpha$ subunit abundances were unchanged in hyperthyroidism was confirmed directly by immuno-blotting by Rapiejko *et al.* (1989); nor was there any significant change in the steady-state level of mRNA for $G_s\alpha$ (Rapiejko *et al.*, 1989). In addition, experiments in which adipocyte G_s from euthyroid and from T_3-treated rats was used to reconstitute regulation of adenylyl cyclase in S49 cyc⁻ cells suggested that hyperthyroidism had not altered the functionality of G_s (Rapiejko and Malbon, 1987). The only study to suggest that hyperthyroidism might cause some modification at the G_s/β-adrenergic receptor level was provided by Malbon (1980a). Treatment of rats with T_3 caused a 2.5-fold decrease in the EC_{50} for displacement of [³H]dihydroalprenolol from adipocyte membranes by isoproterenol, suggesting an increase in the affinity of the β-adrenergic receptor for agonists. Unexpectedly, Gpp(NH)p was totally ineffective in causing any rightwards shift in this displacement curve in hyper-thyroid membranes even though a seven-fold such shift was seen in the euthyroid state. This finding is intriguing but does not appear to have been pursued further.

The situtation in adipocytes from hyperthyroid rats is complicated by the fact that adenylyl cyclase activity is decreased in this state. Rapiejko and Malbon (1987) found decreases of 50%, 35% and 37% respectively in the forskolin-, GTPγS- or fluoride-stimulated activity. However, as mentioned above, cAMP accumulation in response to adrenaline in adipocytes is actually greater in the hyperthyroid state. This apparent anomaly was resolved by Rapiejko and Malbon (1987), who measured cAMP accumulation when adenosine deaminase was added to the cell incubations. Under these experimental conditions cAMP accumulation throughout was greater than in the absence of adenosine deaminase but less accumulation occurred in the hyperthy-roid compared to the euthyroid state. This led Rapiejko and Malbon

(1987) to propose that adipocytes are less sensitive to adenosine in the hyperthyroid state and this was confirmed directly by showing that the EC_{50} for inhibition of cAMP accumulation by PIA was increased three-fold despite a 35% decrease in the B_{max} for binding of cyclohexyl-adenosine to the cell membranes.

On studying [^{32}P]ADP-ribosylation of pertussis toxin substrates in adipocyte membranes, Rapiejko and Malbon (1987) found no significant change in the abundance of 39-kDa or 41-kDa labelled polypeptides in hyperthyroidism. More recently the same laboratory (Ros *et al.*, 1988), using quantitative immunoblotting, have reported that hyperthyroidism causes a 25% decrease in abundance of a 41-kDa polypeptide that cross-reacts with an antiserum against rod transducin and a 20% decrease in a 39-kDa species that was designated as G_o (see Section 2.3 for discussion of this assignment). These small, but significant, decreases in G-protein α subunit abundance were not matched by any change in the levels of 35–36-kDa β subunits (Ros *et al.*, 1988). Quantitation of adipocyte mRNA steady-state levels by Rapiejko *et al.* (1989) showed a 20% decrease in mRNA for $G_i2\alpha$ and a 30% decrease in mRNA for G-protein β subunits.

Table 2. Summary of changes in cell signalling components in rat adipocytes in hyperthyroidism

Component	Comments
β-Adrenergic receptor abundance	Unchanged
$G_s\alpha$ polypeptide abundance	Unchanged
$G_s\alpha$ mRNA abundance	Unchanged
$G_s\alpha$ functionality	Unchanged
Adenylyl cyclase activity	Decreased
Adenosine receptor abundance	Decreased
$G_i\alpha$ polypeptide abundance	Decreased (41 kDa)
$G_o\alpha$ polypeptide abundance	Decreased
$G_i\alpha$ mRNA abundance	Decreased (G_i2)
$G\beta$ polypeptide abundance	Unchanged
$G\beta$ mRNA abundance	Decreased

In summary, a few days of treatment of rats with thyroid hormone (usually T_3) brings about several changes in adipocyte cell signalling (summarized in Table 2). Overall there is enhancement in the lipolytic response of the tissue. When individual components are dissected out there is little evidence for upregulation of the stimulatory pathway;

indeed some downregulation of adenylyl cyclase actually occurs. This appears to be more than compensated for by some downregulation of the inhibitory pathway through which adenosine (and other G_i-coupled agonists) interact with the system. Reduced abundance of 'G_i' contributes to this downregulation.

2.5 Brown adipose tissue

In the acute phase at least, heat generation through uncoupled respiration in brown fat is dependent on noradrenaline-stimulated lipolysis to provide the respiratory substrate. In hypothyroidism there is a severe lesion in thermogenesis (Mory *et al.*, 1981; Seydoux *et al.*, 1982). Although this is partly due to 'downstream' changes at the level of the mitochondrion (Bianco and Silva, 1987; Woodward and Saggerson, 1989), it has also been shown that noradrenaline stimulation of cAMP generation and lipolysis is defective in hypothyroidism (Sundin *et al.*, 1984; Woodward and Saggerson, 1986). As in white adipocytes (Section 2.1), adenosine and its analogues inhibit brown fat cAMP formation and lipolysis (Szillat and Bukowiecki, 1983; Schimmel and McCarthy, 1984) through a pertussis toxin-sensitive pathway (Woodward and Saggerson, 1986). Adenosine deaminase partially restores a normal lipolytic response to noradrenaline in brown adipocytes from hypothyroid rats (Woodward and Saggerson, 1986). Increased sensitivity to adenosine in this state was shown by Woodward and Saggerson (1986), who found a 16-fold decrease in the EC_{50} for inhibition of noradrenaline-stimulated lipolysis by PIA. At present it is not known if this can be attributed to upregulation of 'G_i'.

3 THE CENTRAL NERVOUS SYSTEM (CNS)

3.1 General comments

The fundamental importance of cell signalling in the CNS is reflected by its high abundance of G-proteins. Because of the plethora of receptor–effector interactions and the complexity of cellular and regional organization, the investigations described below are still rather preliminary and it is not yet possible to firmly correlate any changes in G-proteins with altered physiological function.

The need for thyroid hormone for normal development of the brain is beyond question. It is reasonable to speculate that abnormal expression of G-proteins contributes to the severe mental retardation in cretinism. However, there is no experimental evidence at present which can support this conjecture. Comments in this section relate solely to events occurring in the mature brain.

It is frequently implied that the adult brain is insensitive to thyroid hormones. This may be because responses such as increased O_2 consumption or induction of oxidative enzymes which are characteristic of many other tissues are not seen in brain preparations in response to administration of thyroid hormones (Lee and Lardy, 1965; Hemon, 1968; Ismail-Beigi and Edelman, 1971; Schwartz and Oppenheimer, 1978). However, the adult brain contains high-affinity nuclear receptors for thyroid hormones (Oppenheimer *et al.*, 1974; Valcana and Timiras, 1978; Kolodny *et al.*, 1985) as well as significant intracellular contents of T_3 and T_4 (Heninger and Albright, 1975; Obregon *et al.*, 1978).

3.2 Hypothyroidism

Adult hypothyroid humans have a general decrease in brain excitability and numerous specific symptoms indicative of altered cell signalling in the CNS; e.g. cerebellar ataxia, loss of alpha rhythm, increased sensitivity to the depressant actions of phenothiazines and morphine, increased thresholds to sound and light, tiredness, memory impairment, depression, anxiety and occasionally dementia or coma (Reichlin, 1974; Turner and Bagnara, 1976; Utiger, 1981; Gold *et al.*, 1981; Reus, 1986).

Mazurkiewicz and Saggerson (1989) obtained indirect evidence for an increase in 'G_i' in synaptosomal membranes isolated from rat forebrain. In the presence of 100 mM NaCl, inhibition of forskolin-stimulated adenylyl cyclase was not observed in the euthyroid state. By contrast, in membranes from hypothyroid rats appreciable inhibition of the cyclase was seen with 10^{-5} or 10^{-4} M GTP. Mazurkiewicz and Saggerson (1989) also found that inhibition of synaptosomal membrane adenylyl cyclase by PIA was slightly, but significantly, increased in hypothyroidism despite the lack of any significant change in B_{max} or K_D for binding of [^{3}H]PIA to the membranes. Further evidence for alteration at a post-receptor G-protein locus came from study of displacement from synaptosomal membranes of the A_1-adenosine receptor antagonist [^{3}H]diethylphenylxanthine (DPX) by the agonist PIA (Mazurkiewicz and Saggerson, 1989). In the absence of GTP, displacement curves conformed to the presence of two affinity states of the receptor (Hill coefficients were 0.48 and 0.32 respectively in the euthyroid and hypothyroid states). With euthyroid membranes in the presence of 100 μM GTP the displacement curve was shifted rightwards but still displayed two affinity states (Hill coefficient = 0.75). By contrast, in the hypothyroid state addition of 100 μM GTP caused all of the receptors to assume a low-affinity state (Hill coefficient = 0.99).

Orford *et al.* (1991), using anti-peptide sera, provided direct immunological evidence for upregulation of brain G-protein α subunits in hypothyroidism and also showed regional selectivity in these responses (Figure 6). $G_i1α$ abundance was increased two- to three-fold in synaptosomal membranes from the striatum and the cerebral cortex. Smaller non-significant increases in $G_i1α$ were also seen in the cerebellum and hippocampus. By contrast, $G_i1α$ was unchanged in the medulla oblongata and somewhat decreased in the hypothalamus. $G_i2α$ abundance showed four-fold increases in the medulla oblongata and the hippocampus. Elsewhere two- to three-fold increases in abundance of this α subunit were seen in hypothyroidism. Figure 6 shows in the medulla oblongata, the hypothalamus and the hippocampus that hypothyroidism led to substantial increases in the ratio $G_i1α/G_i2α$. Orford *et al.* (1991) also quantitated 'G_o' by immunoblotting and, except in the hypothalamus, found the abundance of the α subunit(s) to be increased in all regions by two- to three-fold in hypothyroidism. The antiserum used in this study did not discriminate the two splice variants $G_o1α$ and $G_o2α$ (Hsu *et al.*, 1990). At present the mechanisms behind these changes remain unknown. Obviously alterations in the amount of individual G-proteins or in the ratios of different α subunits could appreciably alter the sensitivity and scale of neuronal responses to individual neurotransmitters. Alterations in the strengths of different signalling pathways within individual neurones may be expected. However, at this stage our information has been gained by study of rather general anatomical regions of the rat brain. It may be that changes in abundances of G-protein α subunits are much more selectively localized to certain neurones or neuronal tracts within regions. At present there is no information regarding the effect of hypothyroidism on $G_sα$ or on G-protein β subunits.

In the CNS, inhibition of neurotransmitter release is mediated by several types of presynaptic receptors, including A$_1$-adenosine, $α_2$-

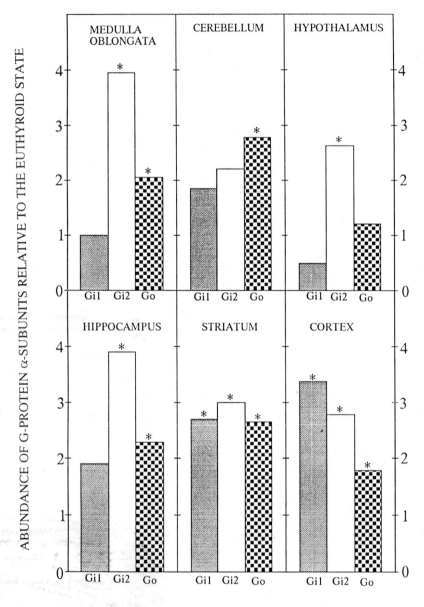

Figure 6. Effect of hypothyroidism on abundances of G-protein α subunits in synaptosomal membranes from regions of the rat brain. The data are from Orford *et al.* (1991). Rats were treated as described in Figure 1. Membranes were obtained from synaptosomes that had been purified on Ficoll gradients. After SDS-PAGE, G-protein α subunits were detected and quantitated using anti-peptide sera. Values for the euthyroid state are set at 1. Statistically significant changes in hypothyroidism are indicated by *.

adrenergic, opiate, $GABA_\beta$, muscarinic cholinergic and dopamine D_2 acting through inhibition of adenylyl cyclase, inhibition of Ca^{2+} channels or increasing K^+ conductance (Proctor and Dunwiddie, 1983; Dolphin *et al.*, 1986; Trussell and Jackson, 1985, 1987; Scott and Dolphin, 1987; Fredholm and Dunwiddie, 1988). Since the pertussis toxin-sensitive G-proteins have been implicated in such responses (Simonds *et al.*, 1989; Brown and Birnbaumer, 1990; Brown, 1990; Birnbaumer, 1990; Dolphin, 1990; Schultz *et al.*, 1990; McKenzie and Milligan, 1990) it is likely that hypothyroidism strengthens signal inputs causing inhibition of neurotransmitter release.

3.3 Hyperthyroidism

Thyroid hormone sensitizes the nervous system. The increased excitability of the CNS after thyroid hormone administration is separable from the effects of the hormone on oxidative metabolism which are seen in most other tissues (Turner and Bagnara, 1976). Hyperthyroid humans exhibit emotional lability, tension, restlessness, agitation and poor concentration (Reichlin, 1974).

At present the only information on G-proteins in this state is from a preliminary study (F.C.L. Leung, M.R. Orford and E.D. Saggerson, unpublished work). In rats treated with T_3 for three days the abundances of $G_i1\alpha$, $G_i2\alpha$ and 'G_o'α were found to be decreased by 63%, 57% and 46% respectively in synaptosomal membranes prepared from the cerebral cortex. By contrast, no such changes were seen in the cerebellum or the medulla oblongata. As an index of loss of 'G' function in the cerebral cortex in hyperthyroidism it was found that GTP no longer increased the Hill coefficient for the displacement from synaptosomal membranes of the A_1-adenosine receptor antagonist [^{3}H]DPCPX by the agonist PIA.

4 THE HEART

4.1 General remarks

It is generally regarded that the effects of thyroid hormones on the heart are very similar to those induced by the catecholamines. Hyperthyroid human subjects exhibit sinus tachycardia and increased myocardial contractility. Atrial fibrillation in approximately 15% of patients is the most important ECG abnormality in thyrotoxicosis.

In addition, minor ST and T wave abnormalities in the ECG are frequently observed in hyperthyroid subjects. In hypothyroidism a number of alterations in cardiac function are found which are essentially opposite to those in hyperthyroidism. Sinus bradycardia is common together with low amplitudes for P, QRS and T waves in the ECG. Arrhythmias are rare in hypothyroidism.

4.2 The G_s pathway

A number of laboratories, using isolated cardiac preparations, have shown that sensitivity to β-adrenergic stimulation is increased in the hyperthyroid state and decreased in hypothyroidism. Hyperthyroid hearts *in vitro* exhibit enhanced β-adrenergic activation of phosphorylase (McNeill and Brody, 1968; Hornbrook and Cabral, 1972), contractility and cAMP-dependent protein kinase (Guarnieri *et al.*, 1980) in response to submaximal levels of isoproterenol. Furthermore, isolated fetal mouse hearts exposed to T_3 in organ culture became more sensitive to the chromotropic effect of β-adrenergic stimulation (Wildenthal, 1972, 1974) and neonatal rat heart cells cultured with T_3 showed enhanced elevation of cAMP in response to adrenaline (Tsai and Chen, 1978). By contrast, Brodde *et al.* (1980) showed in atria from PTU-treated hypothyroid rats that the dose–response curves for increasing force of contraction or elevating cAMP by isoproterenol were moved rightwards.

The pathway for stimulation of adenylyl cyclase through the cardiac β-adrenergic receptor has been dissected to some extent in thyroid disorders. It is generally found that hyperthyroidism increases the abundance of β-adrenergic receptors as quantitated by binding of [3H]dihydroalprenolol to membranes from ventricle or whole heart. Increases in B_{max} of two- to four-fold have been reported (Williams *et al.*, 1977; Tsai and Chen, 1978; Ciaraldi and Marinetti, 1977; Tse *et al.*, 1980; Stiles and Lefkowitz, 1981; Krawietz *et al.*, 1982). By contrast, thyroidectomy decreases the abundance of binding sites for this β-adrenergic antagonist (Banerjee and Kung, 1977; Stiles and Lefkowitz, 1981; Krawietz *et al.*, 1982). Changes in K_D for this ligand were not observed in either thyroid state. When displacement from cardiac membranes of [3H]dihydroalprenolol by the agonist isoproterenol was studied, Stiles and Lefkowitz (1981) found that hyperthyroidism caused a significant four-fold decrease in the K_D for isoproterenol at the high-affinity state of the β-adrenergic receptor. No change in this parameter was seen in hypothyroidism.

No clear picture emerges from the literature concerning the effect of thyroid states on the activity of cardiac adenylyl cyclase. Tse *et al.* (1980) found that T_4 administration to rats had no effect on basal or NaF-stimulated cyclase activity in ventricular homogenates. Krawietz *et al.* (1982) also found little change in the basal activity or in that measured with 100 μM Gpp(NH)p in a crude ventricular preparation from T_3-treated rats. In the same study (Krawietz *et al.*, 1982), thyroid-ectomy was shown to have no effect on the Gpp(NH)p-stimulated cyclase activity but a small increase in basal activity was noted. The general lack of any appreciable effect of thyroid state on ventricular adenylyl cyclase activity was recently confirmed by Levine *et al.* (1990), who found no change in NaF-stimulated or forskolin-stimulated activities in hyperthyroidism or hypothyroidism. By contrast, Brodde *et al.* (1980) assayed basal and NaF-stimulated cyclase activities in atrial membrane fraction from PTU-treated hypothyroid rats and found decreases of approximately 50% in this state.

Krawietz *et al.* (1982) studied the [^{32}P]ADP-ribosylation of 42-kDa and 46-kDa cholera toxin substrates in crude ventricular membranes. The abundance of label was not changed in hyperthyroidism or hypothyroidism. More recently, Levine *et al.* (1990) quantitated the abundance of the 45-kDa $G_s\alpha$ in crude ventricular membranes by immunoblotting. No significant changes were found either after T_4 administration or in PTU-induced hypothyroidism. Accompanying this, Levine *et al.* (1990) also quantitated $G_s\alpha$ mRNA by dot-blot analysis using a cDNA probe. Neither hyperthyroidism nor hypo-thyroidism caused any change in the steady-state level of this mRNA.

In summary, apart from changes in the abundance of the β-adrener-gic receptor there is little change in the G_s pathway in the ventricle that could account for the altered responsiveness of the heart to catecholamines in thyroid states. Without any evidence at this stage, one is led to speculate that altered responsiveness to catecholamines may, at least in hypothyroidism, result from altered sensitivity to agents such as acetylcholine or adenosine that antagonize β-adrenergic effects through pertussis toxin-sensitive pathways (see Chapter 3 for further discussion).

4.3 Changes in 'G_i'

Levine *et al.* (1990) used anti-peptide sera for quantitative immuno-blotting and observed 1.9- and 2.6-fold increases respectively in abun-dances of $G_i2\alpha$ and $G_i3\alpha$ in ventricular membranes from hypothyroid rats. These changes were accompanied by 2.8- and 1.8-fold increases respectively in amounts of the 36-kDa and 35-kDa forms of the G-protein β subunit. By contrast, there were no significant changes in any of these subunits in ventricular membranes from T_4-treated rats. In parallel with the increases in polypeptide abundance there were 2.3-, 1.5-, 1.7- and 2.6-fold increases in the mRNAs corresponding to $G_i2\alpha$, $G_i3\alpha$, $G\beta_{36}$ and $G\beta_{35}$ respectively in hypothyroidism. When Levine et al. (1990) investigated inhibition of forskolin-stimulated adenylyl cyclase by Gpp(NH)p in the ventricular membrane preparations they did not detect any appreciable change in this effect in the hypothyroid state. They interpreted this finding to mean, by contrast with the adipocyte, that 'G_i' is already present at saturating levels in the ventricle even in the euthyroid state. Unfortunately Levine et al. (1990) only tested Gpp(NH)p at one concentration (10 μM) which was already sufficient to inhibit the cyclase activity by approximately 75% in euthyroidism and in both hyper- and hypothyroidism. It remains to be established whether the increase in ventricular 'G_i' in hypothyroidism might be accompanied by enhanced sensitivity to submaximal concentrations of guanine nucleotide. It is now important also to establish whether ventricular adenylyl cyclase in hypothyroidism might be more sensitive to inhibition through receptors that are coupled via 'G_i'. For example, acetylcholine through a muscarinic receptor and adenosine through an A_1 receptor are able to antagonize the ionotropic effect of β-adrenergic agonists in ventricular myocytes. These effects, which are abolished by pertussis toxin, appear to be secondary to inhibition of catecholamine-activated adenylyl cyclase (Schutz et al., 1986; Hescheler et al., 1986; Isenberg et al., 1987). In turn, lowered cAMP level is likely to result in decreased stimulation of both the slow Ca^{2+} inward current (Reuter, 1983; Kameyama et al., 1985; Walsh et al., 1988) and the delayed rectifier K^+ current (Bennett and Begenisich, 1987; Walsh and Kass, 1988; Walsh et al., 1988; Szabo and Otero, 1990) with a resultant reduction in the β-adrenergic ionotropic effect.

In view of the observations that the $G_o\alpha$, 'G_i'α and 'G'β subunits are more abundant in adult rat atrial membranes than in ventricular membranes and that mRNAs for $G_o\alpha$ and $G_i3\alpha$ are more abundant in atria (Luetje et al., 1988), it would seem imperative to extend studies of G-proteins in thyroid states to the atria and to SA node cells. In these regions of the heart acetylcholine muscarinic and A_1-type adenosine receptors appear to couple directly via a pertussis toxin-sensitive G-protein to an inwardly rectifying K^+ channel (Belardinelli and Isenberg, 1983; Endoh et al., 1985; Sorota et al., 1985; Martin et al., 1985; Pfaffinger et al., 1985; West et al., 1987; Isenberg et al., 1987; Szabo and

Otero, 1990). Such studies may help to explain the chronotropic changes that are observed in thyroid disorders.

5 THE LIVER

Unlike in adipose tissue or the heart, hypothyroidism causes substantial upregulation of β-adrenergic regulation of hepatocytes. This manifests as an increased maximal activation of glycogen phosphorylase by isoproterenol after thyroidectomy (Preiksaitis and Kunos, 1979) or PTU administration (Malbon *et al.*, 1978b) and is paralleled by increased cAMP accumulation in cells and increased β-adrenergic activation of adenylyl cyclase in hepatocyte homogenates (Malbon, 1980b). Hypothyroidism similarly enhanced the β-adrenergic stimulation of ureogenesis in rat hepatocytes (Garcia-Sainz *et al.*, 1989). Malbon (1980b) eliminated changes in adenylyl cyclase or cAMP phosphodiesterase activities as possible loci for the enhanced β-adrenergic response on the following grounds. First, hypothyroidism did not appreciably alter either of these enzyme activities. Second, the effect of hypothyroidism was β-adrenergic-specific since responsiveness to glucagon was not upregulated (Malbon *et al.*, 1978b). Malbon (1980b) quantitated abundance of putative β-adrenergic receptors as binding of [^{125}I]iodohydroxybenzylpindolol to hepatocyte membranes and found that hypothyroidism increased the B_{max} by two- to three-fold without any change in the K_D. Malbon (1980b) concluded that this increase in β-adrenergic receptor abundance was the major factor responsible for the enhancement in β-adrenergic stimulation in hypothyroidism. Subsequent work by Garcia-Sainz *et al.* (1989) would seem to support this conclusion. These workers found that PTU-induced hypothyroidism only marginally increased by 17–20% the abundances in hepatocyte membranes of 42-kDa and 46-kDa substrates for [^{32}P]ADP-ribosylation by cholera toxin. In addition, Garcia-Sainz *et al.* (1989) did not observe any significant change in hepatocyte NaF- or forskolin-stimulated adenylyl cyclase activity in hypothyroidism.

Hyperthyroidism due to T_3 administration to rats decreases the maximum effects of isoproterenol or glucagon in elevating cAMP levels in isolated hepatocytes (Malbon and Greenberg, 1982). Both decreased activity of adenylyl cyclase and decreased receptor binding of either hormone could contribute to this downregulation of responses (Malbon and Greenberg, 1982). However, hyperthyroidism did not alter the abundance of hepatocyte membrane substrates that were [^{32}P]ADP-ribosylated by cholera toxin nor did this state alter the ability of Lubrol extracts from hepatocyte membranes to reconstitute Gpp(NH)p or NaF sensitivity of adenylyl cyclase in S49 lymphoma cyc$^-$ cells (Malbon and Greenberg, 1982).

In conclusion, from the studies described there is no evidence for altered abundance or function of G_s in hepatocytes in thyroid states. However, it will probably be worthwhile to attempt confirmation of these conclusions by quantitative immunoblotting. As yet little attention has been paid to hepatocyte pertussis toxin-sensitive signalling pathways in thyroid states. Such studies may prove rewarding since the vasopressin-induced inflow of Ca^{2+} across the hepatocyte plasma membrane is pertussis toxin-sensitive (Hughes *et al.*, 1987) and the activation of hepatocyte glycogen phosphorylase by vasopressin via a Ca^{2+}-mediated pathway is impaired in hypothyroidism (Preiksaitis *et al.*, 1982; Corvera *et al.*, 1984).

6 RETICULOCYTES

Stiles *et al.* (1981) studied the β-adrenergic stimulation of adenylyl cyclase in reticulocyte membranes obtained after phenylhydrazine treatment of rats. Thyroidectomy caused a 50% decrease in the maximum isoproterenol- or NaF-stimulated activities of the cyclase without significant change in the EC_{50} for isoproterenol. Measurement of binding of [^{3}H]dihydroalprenolol to these membranes revealed a 50% decrease in B_{max} without any change in K_D for this β-adrenergic receptor antagonist. When displacement of [^{3}H]dihydroalprenolol by the agonist isoproterenol was examined, Stiles *et al.* (1981) observed a significant decrease in the ratio K_L/K_H in hypothyroidism; i.e. this state increased K_D values for isoproterenol at both a low- and a high-affinity state of the β-adrenergic receptor, but with the larger change seen at the high-affinity site (K_H). Stiles *et al.* (1981) interpreted the increase in K_L/K_H as reflecting a decreased coupling of the receptor to G_s. Stiles *et al.* (1981) quantitated [^{32}P]ADP-ribosylation of the 42-kDa cholera toxin substrate and found this to be decreased by 40% in the hypothyroid state.

In summary, it appears in the abnormal situation of reticulocytosis that concomitant hypothyroidism decreases all three components of the β-adrenergic receptor–G_s–cyclase pathway by 40–50%. The consequences of these changes for reticulocyte maturation are unclear.

7 CONCLUSION

In terms of changes in G-protein subunit abundance I have described some instances that involve the β subunits and various pertussis toxin-sensitive α subunits. From studies of adipose tissue, heart and the CNS it emerges that the usual, but not universal, rule is that hypo-thyroidism increases abundance of these components. Where studied, changes in 'G$_i$' in hyperthyroid states were smaller and less signifi-cant. It is not clear at present whether this reflects the relatively short periods of T$_3$ or T$_4$ treatment that have been used. So far, unlike in insulin-dependent diabetes (Bushfield *et al.*, 1990a; Strassheim *et al.*, 1990) or genetic obesity (Bushfield *et al.*, 1990b), there is no evidence that thyroid state causes any changes in the functionality of 'G$_i$' in addition to the reported alterations in abundance. Unlike 'G$_i$' and G$_o$, in the cases so far examined G$_s$ appears to be constitutive in terminally differentiated cells. The only deviation from this was the reticulocyte but this may prove an exception both because of the abnormal situation (phenylhydrazine treatment of the donor animals) and because this is a cell still in the process of differentiation/maturation.

For the future there are other systems that must merit investigation in thyroid states, e.g. the non-cardiac elements of the vascular system, skeletal muscle types, the peripheral nervous system. In addition, changes in G-proteins in the CNS and the heart should be studied at finer levels of anatomical resolution. Finally, efforts will be made to understand the mechanisms underlying altered expression of G-protein subunits in thyroid disorders.

REFERENCES

Armstrong, K.J., Stouffer, J.E., Van Inwegen, R.G., Thompson, W.J. and Robison, G.A. (1974). *J. Biol. Chem.* **249**, 4226–4231.

Banerjee, S.P. and Kung, L.S. (1977). *Eur. J. Pharmacol.* **43**, 207–208.

Beebe, S.J., Redmon, J.B., Blackmore, P.F. and Corbin, J.D. (1985). *J. Biol. Chem.* **260**, 15781–15788.

Belardinelli, L. and Isenberg, G. (1983). *Am. J. Physiol.* **244**, H734–H737.

Bennett, P.B. and Begenisich, T.B. (1987). *Pflugers Arch.* **410**, 217–219.

Bianco, A.C. and Silva, J.E. (1987). *J. Clin. Invest.* **79**, 295–300.

Bilous, R.W. and Tunbridge, W.M.G. (1988). In: *Clinical Endocrinology and Metabolism*, Vol. 2 (Lazarus, J.H. and Hall, R., eds), pp. 531–540. Baillière Tindall, London.

Birnbaumer, L. (1990). *Annu. Rev. Pharmacol. Toxicol.* **30**, 675–705.

Bray, G.A. and Goodman, H.M. (1965). *Endocrinology (Baltimore)* **76**, 323–328.

Brodde, O-E., Schumann, H-J. and Wagner, J. (1980). *Mol. Pharmacol.* **17**, 180–186.

Brown, A.M. and Birnbaumer, L. (1990). *Annu. Rev. Physiol.* **52**, 197–213.

Brown, D.A. (1990). *Annu. Rev. Physiol.* **52**, 215–242.

Bushfield, M., Griffiths, S.L., Murphy, G.J., Pyne, N.J., Knowler, J.T., Milligan, G., Parker, P.J., Mollner, S. and Houslay, M.D. (1990a). *Biochem. J.* **271**, 365–372.

Bushfield, M., Pyne, N.J. and Houslay, M.D. (1990b). *Eur. J. Biochem.* **192**, 537–542.

Caldwell, A. and Fain, J.N. (1971). *Endocrinology (Baltimore)* **89**, 1195–1204.

Chohan, P., Carpenter, C. and Saggerson, E.D. (1984). *Biochem. J.* **223**, 53–59.

Ciaraldi, T. and Marinetti, G.V. (1977). *Biochem. Biophys. Res. Commun.* **74**, 984–991.

Correze, C., Laudat, M.H., Laudat, P. and Nunez, J. (1974). *Mol. Cell. Endocrinol.* **1**, 309–327.

Corvera, S., Hernandez-Sotomayor, S.M.T. and Garcia-Sainz, J.A. (1984). *Biochim. Biophys. Acta* **803**, 95–105.

Debons, A.F. and Schwartz, I.L. (1961). *J. Lipid Res.* **2**, 86–89.

Degerman, E., Smith, C.J., Tornqvist, H., Vasta, V., Belfrage, P. and Manganiello, V.C. (1990). *Proc. Natl. Acad. Sci. USA* **87**, 533–537.

Deykin D. and Vaughan, M. (1963). *J. Lipid Res.* **4**, 200–203.

Dolphin, A.C. (1990). *Annu. Rev. Physiol.* **52**, 243–255.

Dolphin, A.C., Forda, S.R. and Scott, R.H. (1986). *J. Physiol. (Lond.)* **373**, 47–61.

Ebert, R. and Schwabe, U. (1973). *Naunyn Schmiedeberg's Arch. Pharmacol.* **278**, 247–259.

Elks, M.L., Manganiello, V.C. and Vaughan, M. (1983). *J. Biol. Chem.* **258**, 8582–8587.

Endoh, M., Maruyama, M. and Iijima, T. (1985). *Am. J. Physiol.* **249**, H309–H320.

Fain, J.N. (1973). *Mol. Pharmacol.* **9**, 595–604.

Fain, J.N. and Wieser, P.B. (1975). *J. Biol. Chem.* **250**, 1027–1034.

Fernandez, B.M. and Saggerson, E.D. (1978). *Biochem. J.* **174**, 111–118.

Fisher, J.N. and Ball, E.G. (1967). *Biochemistry* **6**, 637–647.

Fredholm, B.B. and Dunwiddie, T.V. (1988). *Trends Pharmacol. Sci.* **9**, 130–134.

Garcia-Sainz, J.A. and Medina-Martinez, O. (1989). *Biochem. Int.* **19**, 899–907.

Garcia-Sainz, J.A., Huerta-Bahena M.E. and Malbon, C.C. (1989). *Am. J. Physiol.* **256**, C384–C389.

Giudicelli, Y. (1978). *Biochem. J.* **176**, 1007–1010.

Gold, M.S., Pottash, A.L.C. and Extein, I. (1981). *J. Am. Med. Assoc.* **245**, 1919–1922.

Goodman, H.M. and Knobil, E. (1959). *Proc. Soc. Exp. Biol. Med.* **100**, 195–197.

Goswami, A. and Rosenberg, I.N. (1978). *Endocrinology (Baltimore)* **103**, 2223–2233.

Green, A. (1983). *FEBS Lett.* **152**, 261–264.

Green, A. (1987). *J. Biol. Chem.* **262**, 15702–15707.

Green, A., Johnson, J.L. and Milligan, G. (1990). *J. Biol. Chem.* **265**, 5206–5210.

Guarnieri, T., Filburn, C.R. and Beard, E.S. (1980). *J. Clin. Invest.* **65**, 861–868.

Hemon, P. (1968). *Biochim. Biophys. Acta* **151**, 681–683.

Heniger, R.W. and Albright, E.C. (1975). *Proc. Soc. Exp. Biol. Med.* **150**, 137–142.

Hescheler, J., Kameyama, M. and Trautwein, W. (1986). *Pflugers Arch.* **407**, 182–189.

Hinsch, K.D., Rosenthal, W., Spicher, K., Binder, T., Gausepohl, H., Frank, R., Schultz, G. and Joost, H.G. (1988). *FEBS Lett.* **238**, 191–196.

Honeyman, T.W., Strohsnitter W., Schied, C.R. and Schimmel, R.J. (1983). *Biochem. J.* **212**, 489–498.

Hornbrook, K.R. and Cabral, A. (1972). *Biochem. Pharmacol.* **21**, 897–907.

Hsu, W.H., Rudolph, U., Sanford, J., Bertrand, P., Olate J., Nelson, C., Moss, L.G., Boyd, A.E., Codina, J. and Birnbaumer, L. (1990). *J. Biol. Chem.* **265**, 11220–11226.

Hughes, B.P., Crofts, J.N., Auld, M.A., Read, L.C. and Barritt, G.J. (1987). *Biochem. J.* **248**, 911–918.

Ichikawa, A., Matsumoto, H., Sakato, N. and Tomita, K. (1971). *J. Biol. (Tokyo)* **69**, 1055–1064.

Isenberg, G., Cerbai, E. and Klöckner, U. (1987). In: *Topics and Perspectives in Adenosine Research* (Gerlach, E. and Becker, B.F., eds), pp. 323–335. Springer-Verlag, Berlin, Heidelberg, New York, London and Paris.

Ismail-Beigi, F. and Edelman, I.S. (1971). *J. Gen. Physiol.* **57**, 710–722.

Jamal, Z. and Saggerson, E.D. (1987). *Biochem. J.* **245**, 881–886.

Joost, H.G. and Steinfelder, H.J. (1982). *Mol. Pharmacol.* **22**, 614–618.

Kemeyama, M., Hofmann, F. and Trautwein, W. (1985). *Pflugers Arch.* **405**, 285–293.

Kolodny, J.M., Larsen, P.R. and Silva, J.E. (1985). *Endocrinology (Baltimore)* **116**, 2019–2028.

Krawietz, W., Werdan, K. and Erdmann, E. (1982). *Biochem. Pharmacol.* **31**, 2463–2469.

Krishna, G., Hynie, S. and Brodie, B.B. (1968). *Proc. Natl. Acad. Sci. USA* **59**, 884–889.

Lee, Y-P. and Lardy, H.A. (1965). *J. Biol. Chem.* **240**, 1427–1437.

Levine, M.A., Feldman, A.M., Robishaw, J.D., Ladenson, P.W., Ahn, T.G., Moroney, J.F. and Smallwood, P.M. (1990). *J. Biol. Chem.* **265**, 3553–3560.

Londos, C., Cooper, D.M.F. and Rodbell, M. (1981). *Adv. Cyclic Nucleotide Res.* **14**, 163–187.

Luetje, C.W., Tietje, K.M., Christian, J.L. and Nathanson, N.M. (1988). *J. Biol. Chem.* **263**, 13357–13365.

Malbon, C.C. (1980a). *Mol. Pharmacol.* **18**, 193–198.

Malbon, C.C. (1980b). *J. Biol. Chem.* **255**, 8692–8699.

Malbon, C.C. and Gill, D.M. (1979). *Biochim. Biophys. Acta* **586**, 518–527.

Malbon, C.C. and Graziano, M.P. (1983). *FEBS Lett.* **155**, 35–38.

Malbon, C.C. and Greenberg, M.L. (1982). *J. Clin. Invest.* **69**, 414–426.

Malbon, C.C., Moreno, F.J., Cabelli, R.J. and Fain, J.N. (1978a). *J. Biol. Chem.* **253**, 671–678.

Malbon, C.C., Li, S-Y. and Fain, J.N. (1978b). *J. Biol. Chem.* **253**, 8820–8825.

Malbon, C.C., Graziano, M.P. and Johnson, G.L. (1984). *J. Biol. Chem.* **259**, 3254–3260.

Malbon, C.C., Rapiejko, P.J. and Mangano, T.J. (1985). *J. Biol. Chem.* **260**, 2558–2564.

Martin, J.M., Hunter, D.D. and Nathanson, N.M. (1985). *Biochemistry* **24**, 7521–7525.

Mazurkiewicz, D. and Saggerson, E.D. (1989). *Biochem. J.* **263**, 829–835.

McKenzie, F.R. and Milligan, G. (1990). *Biochem. J.* **267**, 391–398.

McNeill, J.H. and Brody, T.M. (1968). *J. Pharmacol. Exp. Ther.* **161**, 40–46.

Milligan, G. and Saggerson, E.D. (1990). *Biochem. J.* **270**, 765–769.

Milligan, G., Spiegel, A.M., Unson, C.G. and Saggerson, E.D. (1987). *Biochem. J.* **247**, 223–227.

Mills, I., Garcia-Sainz, J.A. and Fain, J.N. (1986). *Biochim. Biophys. Acta* **876**, 619–630.

Mitchell, F.M., Griffiths, S.L., Saggerson, E.D., Houslay, M.D., Knowler, J.T. and Milligan, G. (1989). *Biochem. J.* **262**, 403–408.

Moreno, F.J., Mills, I., Garcia-Sainz, J.A. and Fain, J.N. (1983). *J. Biol. Chem.* **258**, 10938–10943.

Mory, G., Ricquier, D., Pesquies, P. and Hemon, P. (1981). *J. Endocrinol.* **91**, 515–524.

Murayama, T. and Ui, M. (1983). *J. Biol. Chem.* **258**, 3319–3326.

Obregon, M.J., Morreale de Escobar, G. and Escobar del Rey, F. (1978). *Endocrinology (Baltimore)* **103**, 2145–2156.

Ohisalo, J.J. and Milligan, G. (1989). *Biochem. J.* **260**, 843–847.

Ohisalo, J.J. and Stouffer, J.E. (1979). *Biochem. J.* **178**, 249–251.

Ohisalo, J.J., Strandberg, H., Kostiainen, E., Kuusi, T. and Ehnholm, C. (1981). *FEBS Lett.* **132**, 121–123.

Ohisalo, J.J., Stoneham, S. and Keso, L. (1987). *Biochem. J.* **246**, 555–557.

Olansky, L., Myers, G.A., Pohl, S.L. and Hewlett, E.L. (1983). *Proc. Natl. Acad. Sci. USA* **80**, 6547–6551.

Olsson, H. and Belfrage, P. (1987). *Eur. J. Biochem.* **168**, 399–405.

Oppenheimer, J.H., Schwartz, H.L. and Surks, M.I. (1974). *Endocrinology (Baltimore)* **95**, 879–902.

Orford, M., Mazurkiewicz, D., Milligan, G.and Saggerson, E.D. (1991). *Biochem. J.* **275**, 183–186.

Parsons, W.J. and Stiles, G.L. (1987). *J. Biol. Chem.* **262**, 841–847.

Pfaffinger, P.J., Martin, J.M., Hunter, D.D., Nathanson, N.M. and Hille, B. (1985). *Nature (Lond.)* **317**, 536–538.

Preiksaitis, H.G. and Kunos, G. (1979). *Life Sci.* **24**, 42–43.

Preiksaitis, H.G., Kan, W.H. and Kunos, G. (1982). *J. Biol. Chem.* **257**, 4321–4327.

Proctor, W.R. and Dunwiddie, T.V. (1983). *Neurosci. Lett.* **35**, 197–201.

Rapiejko, P.J. and Malbon, C.C. (1987). *Biochem. J.* **241**, 765–771.

Rapiejko, P.J., Northup, J.K., Evans, T., Brown, J.E. and Malbon, C.C. (1986). *Biochem. J.* **240**, 35–40.
Rapiejko, P.J., Watkins, D.C., Ros, M. and Malbon, C.C. (1989). *J. Biol. Chem.* **264**, 16183–16189.
Rebourcet, M-C., Carpéné, C. and Lavau, M. (1988). *Biochem. J.* **252**, 679–682.
Reichlin, S. (1974). In: *Textbook of Endocrinology*, 5th edn (Williams, R.H., ed.), pp. 774–831. W.B. Saunders Co., Philadelphia, London, New York.
Reus, V.I. (1986). *Annu. Rev. Med.* **37**, 205–214.
Reuter, H. (1983). *Nature (Lond.)* **301**, 569–574.
Rich, C., Bierman, E.L. and Schwartz, I.L. (1959). *J. Clin. Invest.* **38**, 275–278.
Rodbell, M. (1964). *J. Biol. Chem.* **239**, 375–380.
Ros, M., Northup, J.K. and Malbon, C.C. (1988). *J. Biol. Chem.* **263**, 4362–4368.
Rouot, B., Carrette, J., Lafontan, M., Lan Tran, P., Fehrentz, J-A., Bockaert, J. and Toutant, M. (1989). *Biochem. J.* **260**, 307–310.
Saggerson, E.D. (1980). *FEBS Lett.* **115**, 127–128.
Saggerson, E.D. (1985). In: *New Perspectives in Adipose Tissue: Structure, Function and Development* (Cryer, A. and Van, R.L.R., eds), pp. 87–120. Butterworths, London.
Saggerson, E.D. (1986). *Biochem. J.* **238**, 387–394.
Saggerson, E.D. (1988). In: *Phosphatidate Phosphohydrolase*, Vol. 1 (Brindley, D.N., ed.), pp. 79–124. CRC Press, Inc., Boca Raton, Florida.
Schimmel, R.J. and McCarthy, L. (1984). *Am. J. Physiol.* **246**, C301–C307.
Schultz, G., Rosenthal, W., Hescheler, J. and Trautwein, W. (1990). *Annu. Rev. Physiol.* **52**, 275–292.
Schutz, W., Freissmuth, M., Hausleithner, V. and Tuisl, E. (1986). *Naunyn-Schmiedeberg's Arch. Pharmacol.* **333**, 156–162.
Schwabe, U., Ebert, R. and Erbler, H.C. (1973). *Naunyn-Schmiedeberg's Arch. Pharmacol.* **276**, 133–148.
Schwabe, U., Ebert, R. and Erbler, H.C. (1975). *Adv. Cyclic Nucleotide Res.* **5**, 569–584.
Schwartz, H.L. and Oppenheimer, J.H. (1978). *Endocrinology (Baltimore)* **103**, 943–948.
Scott, R.H. and Dolphin, A.C. (1987). In: *Topics and Perspectives in Adenosine Research* (Gerlach, E. and Becker, B.F., eds), pp. 549–558. Springer-Verlag, Berlin, Heidelberg, New York, London, Paris.
Seydoux, J., Giacobino, J.P. and Girardier, L. (1982). *Mol. Cell. Endocrinol.* **25**, 213–226.
Simonds, W.F., Goldsmith, P.K., Unson, C.G. and Spiegel, A.M. (1989). *Proc. Natl. Acad. Sci. USA* **86**, 7809–7813.
Smith, U., Kuroda, M. and Simpson, I.A. (1984). *J. Biol. Chem.* **259**, 8758–8763.
Sorota, S., Tsuji, Y., Tajima, T. and Pappano, A.J. (1985). *Circ. Res.* **57**, 748–758.
Stiles, G.L. and Lefkowitz, R.J. (1981). *Life. Sci.* **28**, 2529–2536.
Stiles, G.L., Stadel, J.M., De Lean, A. and Lefkowitz, R.J. (1981). *J. Clin. Invest.* **68**, 1450–1455.
Stralfors, P. and Belfrage, P. (1983). *J. Biol. Chem.* **258**, 15146–15152.
Stralfors, P. and Honnor, R.C. (1989). *Eur. J. Biochem.* **182**, 379–385.

Stralfors, P., Björgell, P. and Belfrage, P. (1984). *Proc. Natl. Acad. Sci. USA* **81**, 3317–3321.

Strassheim, D., Milligan, G. and Houslay, M.D. (1990). *Biochem. J.* **266**, 521–526.

Sundin, U., Mills, I. and Fain, J.N. (1984). *Metab. Clin. Exp.* **33**, 1028–1033.

Szabo, G. and Otero, A.S. (1990). *Annu. Rev. Physiol.* **52**, 293–305.

Szillat, D. and Bukowiecki, L.J. (1983). *Am. J. Physiol.* **245**, E555–E559.

Trussell, L.O. and Jackson, M.B. (1985). *Proc. Natl. Acad. Sci. USA* **82**, 4857–4861.

Trussell, L. O. and Jackson, M.B. (1987). *J. Neurosci.* **7**, 3306–3316.

Tsai, J.S. and Chen. A. (1978). *Nature (Lond.)* **275**, 138–140.

Tse, J., Wrenn, R.W. and Kuo, J.F. (1980). *Endocrinology* **107**, 6–12.

Tunbridge, W.M.G. (1978). In: *The Thyroid a Fundamental and Clinical Text*, 5th edn (Ingbar, S.H. and Braverman, L.E., eds), pp. 625–635. J.B. Lippincott Company, Philadelphia.

Turner, C.D. and Bagnara, J.T. (1976). *General Endocrinology*, 6th edn. pp. 178–224. W.B. Saunders Co., Philadelphia, London, New York.

Utiger, R.D. (1981). In: *Endocrinology and Metabolism* (Felig, P., Baxter, J.D., Broadus, A.E. and Frohman, L.A., eds), pp. 389–472. McGraw-Hill, New York.

Valcana, T. and Timiras, P.S. (1978). *Mol. Cell. Endocrinol.* **11**, 31–41.

Van Inwegen, R.G., Bobison, G.A., Thompson, W.J., Armstrong, K.J. and Stouffer, J.E. (1975). *J. Biol. Chem.* **250**, 2452–2456.

Walsh, K.B. and Kass, R.S. (1988). *Science* **242**, 67–69.

Walsh, K.B., Begenisich, T.B. and Kass, R.S. (1988). *Pflugers Arch.* **411**, 232–234.

West, G.A., Giles, W. and Belardinelli, L. (1987). In: *Topics and Perspectives in Adenosine Research* (Gerlach, E. and Becker, B.F., eds), pp. 336–343. Springer-Verlag, Berlin, Heidelberg, New York, London, Paris.

Wildenthal, K. (1972). *J. Clin. Invest.* **51**, 2702–2709.

Wildenthal, K. (1974). *J. Pharmacol. Exp. Ther.* **190**, 272–279.

Williams, L.T., Lefkowitz, R.L., Watanabe, A.M., Hathaway, D.R. and Besch, H.R. (1977). *J. Biol. Chem.* **252**, 2787–2789.

Wong, E.H.A., Ooi, S., Loten, E.G. and Sneyd, J.G.T. (1985). *Biochem. J.* **227**, 815–821.

Woodward, J.A. and Saggerson, E.D. (1986). *Biochem. J.* **238**, 395–403.

Woodward, J.A. and Saggerson, E.D. (1989). *Biochem. J.* **263**, 341–345.

CHAPTER EIGHT

Alcoholism: A Possible G-protein Disorder

ADRIENNE S. GORDON,*†‡ DARIA MOCHLY-ROSEN†*‡ and IVAN DIAMOND*§‡
* The Ernest Gallo Clinic and Research Center, San Francisco General Hospital, San Francisco, CA 94110, USA
† Department of Neurology, University of California, San Francisco, California, USA
‡ Department of Pharmacology, University of California, San Francisco, California, USA
§ Department of Pediatrics, University of California, San Francisco, California, USA

1 CLINICAL ASPECTS

1.1 Alcoholism

Alcoholics are addicted to alcohol, craving and consuming alcoholic beverages without apparent satiation; they are remarkably tolerant to the intoxicating and depressant effects of ethanol. When drinking is discontinued, alcoholics develop many neurological abnormalities that reflect hyperexcitability of the central and peripheral nervous system. This is easily recognized as an alcohol withdrawal syndrome (see below), and considered to be evidence of physical dependence on ethanol.

1.2 Genetic factors in alcoholism

Socio-economic and psychological factors influence drinking habits and the prevalence of alcohol abuse and alcoholism. However, there is recent persuasive evidence that genetic factors play a role in the development of alcoholism (Kiianmaa et al., 1989). In addition to the occurrence of alcoholism in families, studies of identical twins, alcoholic parents and offspring, and adopted children from alcoholic parents, consistently suggest a genetically transmitted susceptibility for alcoholism (Cloninger, 1987; Devor and Cloninger, 1989; Goodwin, 1987). Cloninger and colleagues have identified a type of 'male-limited' alcoholism in fathers and sons with anti-social, impulsive, novelty-seeking behaviour who become alcoholics in teenage years; such alcoholics appear to be unable to abstain from drinking throughout life. When they studied children who were raised in foster homes, they found that this type of alcoholism in the biological father was a much greater predictor for alcoholism in the son than the environment in which the boy was raised (Cloninger, 1987).

1.3 Clinical pharmacology of ethanol

Ethanol is rapidly and completely absorbed from the gastrointestinal tract into the circulation within minutes after drinking (Goldstein, 1983). It is then widely distributed to all organs and fluid compartments in the body, readily equilibrating into total body water and intercalating into biological membranes. Ninety to ninety-eight per cent of ethanol is removed by metabolism in liver and the remainder is excreted by the kidneys, lungs and skin. An important rate-limiting step in ethanol metabolism is oxidation to acetaldehyde by alcohol dehydrogenase in the liver. Acetaldehyde is then converted to acetate by aldehyde dehydrogenase. There is very little alcohol dehydrogenase activity in brain; ethanol itself appears to be responsible for most of the behavioural and neurological consequences of drinking.

1.4 Ethanol intoxication

There is virtually no blood–brain barrier to ethanol and uptake into the brain is limited primarily by cerebral blood flow and capillary perfusion. Therefore, within a short period of time after drinking, the concentration of ethanol in the brain is nearly identical to the level of ethanol in the blood. In non-alcoholics, rising blood alcohol levels of 50–150 mg/dl (10–30 mM) are associated with symptoms of intoxication. Symptoms vary directly with the rate of drinking and are more severe when the blood alcohol concentration is rising than when it is falling (Goldstein, 1983). Most individuals feel euphoric, lose social inhibitions and manifest expansive, sometimes garrulous, behaviour, while others may become gloomy, belligerent, or even explosively combative. Some people do not experience euphoria but instead become sleepy after moderate drinking; they rarely abuse alcohol. With continued drinking, cerebellar and vestibular function deteriorates and at higher blood ethanol levels the findings of central nervous system depression predominate. Progressive lethargy leads to hyporeflexia, hypotension, depression of the brainstem respiratory centre, and coma without focal signs. In non-alcoholics, fatalities may occur at 500 mg/dl, usually because of respiratory depression with ventilatory acidosis and hypotension. Alcoholics are more resistant to the acute intoxicating effects of ethanol than non-alcoholics (Johnson *et al.*, 1982; Lindblad and Olsson, 1976; Urso *et al.*, 1981; Watanabe *et al.*, 1985).

1.5 Acute and chronic tolerance to ethanol

Signs of intoxication may not always be present when blood ethanol concentrations are elevated. Moreover, a slow rise in blood ethanol levels produces less intoxication than a more rapid rise to the same concentrations. Tolerance to ethanol may develop during a single bout of drinking (Goldstein, 1983; Victor and Adams, 1953) and is characterized by a reduced intoxicating response to ethanol. Indeed, subjects can become sober at blood ethanol levels higher than those at which intoxication first developed (Mirsky *et al.*, 1941). This phenomenon is known as 'acute tolerance' and appears to be due primarily to short-term adaptive changes in the central nervous system.

The magnitude of increased tolerance to ethanol in chronic alcoholics is also not often appreciated. Chronic alcoholics have increased resistance to the intoxicating effects of ethanol, and can appear to be sober at blood alcohol levels of 400–500 mg/dl (Urso *et al.*, 1981), concentrations which are severely intoxicating or even fatal in naive individuals. This is known as 'chronic tolerance'. Indeed, a serum ethanol level of 1510 mg/dl was documented in an ambulatory chronic alcoholic who had stopped drinking three days earlier (Lindblad and Olsson, 1976). Therefore, despite legal definitions of intoxication at blood alcohol levels above 80–100 mg/dl, a single blood ethanol determination may not accurately measure the extent of drunkenness (Urso *et al.*, 1981).

1.6 Alcohol withdrawal syndrome

Ethanol is a central nervous system depressant. The nervous system appears to adapt to chronic ethanol exposure by increasing neural mechanisms that counteract the depressant effects of alcohol. When drinking is abruptly reduced or discontinued, these adaptive changes appear to persist unopposed by ethanol, and a hyperexcitable 'ethanol withdrawal syndrome' develops (Porter *et al.*, 1991). This is considered to be evidence of 'physical dependence' on ethanol. The clinical features of the ethanol withdrawal syndrome consist of several characteristic abnormalities which vary in severity (Victor and Adams, 1953). These include tremulousness, disordered perceptions, convulsions, and delirium tremens. The alarming symptoms of ethanol withdrawal are best managed by substituting another central nervous system depressant. However, alcoholics undergoing withdrawal are very resistant to sedatives ('cross-tolerance') and large doses are often required to calm these agitated patients.

2 CELLULAR MODELS OF ETHANOL INTOXICATION, TOLERANCE, AND PHYSICAL DEPENDENCE

Current evidence suggests that acute intoxication, tolerance, and physical dependence are related to acute and chronic ethanol-induced changes in the molecular properties and function of neural membranes which lead to altered signal transduction. Ethanol is readily taken up by cell membranes (Rottenberg, 1986), causing acute increases in membrane fluidity (Goldstein and Chin, 1981). After prolonged exposure to ethanol, however, cells exhibit tolerance to the acute fluidizing effects of ethanol (Goldstein, 1986), apparently because of ethanol-induced changes in membrane lipids (Sun and Sun, 1985), including phosphatidylinositol (Taraschi *et al.*, 1986). However, it is unlikely that ethanol-induced changes in membrane fluidity could be responsible for altered signal transduction since other lipid-fluidizing agents did not produce the same effects as ethanol (Luthin and Tabakoff, 1984; Rabin *et al.*, 1986). As will be discussed in this chapter, acute exposure to ethanol increases cAMP accumulation in many different cell types. However, this ethanol-induced increase in cAMP accumulation was not correlated with changes in membrane fluidity produced by ethanol (Rabin *et al.*, 1986). Instead, membrane proteins responsible for altered signal transduction may be directly affected by ethanol, or ethanol may cause changes in the specific lipid or microenvironment required for optimal function of specific protein components of signal transduction systems (Sun and Sun, 1985). Ethanol has been found to alter components of several signal transduction systems including neurotransmitter receptors (Charness *et al.*, 1986; Mhatre and Ticku, 1989; Valverius *et al.*, 1989; White *et al.*, 1990), ion channels (Daniell and Leslie, 1986; Farrar *et al.*, 1989; Messing *et al.*, 1986, Reynolds *et al.*, 1990), second messenger-generating enzymes (Diamond *et al.*, 1987; Gordon *et al.*, 1986; Hoffman and Tabakoff, 1986; Richelson *et al.*, 1986; Stenstrom *et al.*, 1986), G-proteins (Mochly-Rosen *et al.*, 1988; Charness *et al.*, 1988; Saito *et al.*, 1987), neuromodulators (Nagy *et al.*, 1989, 1990), and gene expression (Miles and Sturdivant, 1991). Many of these systems which are acutely affected by ethanol become tolerant, which is manifested by a decreased response to ethanol on rechallenge with ethanol. We will review available data on the effects of ethanol on the components of the cAMP signal transduction system, including receptors, GTP-binding proteins, and adenylyl cyclase, in brain, cultured cell lines and human cells. This system is the most extensively studied and considerable progress has been made in understanding the interaction of ethanol with its components. Moreover, as will be discussed in the following, alterations in the cAMP signal transduction system have been found in both freshly isolated and cultured cells from alcoholics, suggesting that alterations in the cAMP signal transduction system may be involved in the pathophysiology of alcoholism.

2.1 Brain

We will first review the literature on the effects of *in vivo* ethanol exposure on the cAMP signal transduction system in brain. The literature is inconsistent due to variation in route and quantity of ethanol administration, heterogeneity of cell types present in the selected tissue, inability to distinguish between direct effects of ethanol and secondary trans-synaptic or hormonal effects, and the need for cell disruption before measuring cAMP levels. We present a sampling of this data to provide a historical perspective on the importance of model systems in elucidating the molecular mechanisms underlying the effects of ethanol on signal transduction.

2.1.1 cAMP

Early studies on the effects of ethanol on the cAMP signal transduction system focused on alterations in cAMP levels in various brain regions after treatment with ethanol by feeding, injection, or inspiration of ethanol vapour. For example, 1.5 h after intraperitoneal injection of rats with ethanol, Ferko *et al.* observed a 25% reduction in cAMP levels in the striatum, but not in cortex, cerebellum, or hypothalamus. After 48 h of ethanol administration, cAMP levels returned to control levels (Ferko *et al.*, 1982), indicating a transient effect of ethanol. In contrast, gastric intubation of ethanol caused a 40% decrease in cAMP levels in all areas of the brain after 3 h of ethanol administration; this decrease was sustained until ethanol was withdrawn and cAMP levels returned to normal (Shen *et al.*, 1983).

Ethanol-induced alterations in cAMP levels have also been estimated by changes in cAMP-dependent phosphorylation of brain membrane proteins. Acute exposure to ethanol decreased phosphorylation of several cAMP-dependent protein kinase substrates in the striatum, including DARPP-32, a 32-kDa protein whose phosphorylation is known to be regulated by cAMP. Chronic ingestion of ethanol also resulted in inhibition of protein phosphorylation in striatal preparations (Rius *et al.*, 1987). Subsequent studies in brain slices, homogenates and synaptosomal membranes focused on the identification of the component(s) of the cAMP signal transduction system that are altered by ethanol.

Ethanol has little or no effect on G_i-mediated inhibition of cAMP accumulation (Hoffman and Tabakoff, 1986; Tabakoff and Hoffman, 1983; Rabin, 1985; Hoffman *et al.*, 1987). Therefore, this chapter is focused on the effects of ethanol on stimulation of intracellular cAMP levels by receptors coupled to adenylyl cyclase via G_s.

2.1.2 Adenylyl cyclase

Adenylyl cyclase has been proposed as a specific target for ethanol (Rabin and Molinoff, 1981; Luthin and Tabakoff, 1984; Hoffman *et al.*, 1982). However, in most cases, the assay conditions for adenylyl cyclase do not eliminate possible effects of ethanol on G_s and its coupling to this enzyme. Saito *et al.* (1985) reported that adenylyl cyclase is directly affected by acute treatment with ethanol. They showed that ethanol caused a dose-dependent increase in cAMP production in a digitonin extract of mouse cortical membranes. Since adenylyl cyclase activity in this preparation was insensitive to GTP and fluoride, and therefore uncoupled to G_s, ethanol appeared to stimulate adenylyl cyclase directly. However, Luthin and Tabakoff (1984) found that ethanol had little effect on basal adenylyl cyclase activity in mouse striatal membranes. The greatest effect of ethanol was observed in the presence of guanine nucleotides, suggesting that receptor-activated coupling of G_s to adenylyl cyclase is the principal target of ethanol.

2.1.3 Receptor-stimulated cAMP production

Acute treatment with ethanol caused an increase in dopamine receptor-stimulated cAMP production in mouse striatal membranes (Luthin and Tabakoff, 1984; Rabin and Molinoff, 1981). This increase was not due to release of endogenous neurotransmitters and was reversed on ethanol withdrawal. Isoproterenol-stimulated cAMP levels in homogenates of cerebral cortex and cerebellum were also increased by ethanol (Rabin and Molinoff, 1981). In contrast, Okuda *et al.* found that noradrenaline receptor-stimulated cAMP levels in rat cortical slices were not affected by acute exposure to ethanol (Okuda *et al.*, 1984). Chronic exposure to ethanol produced different results in various brain regions. In brain homogenates from both rat and mouse, isoproterenol and noradrenaline-stimulated cAMP levels were increased (Smith, 1981; Smith *et al.*, 1981). Similarly, in slices of rat striatum, chronic ethanol exposure increased PGE_1-stimulated cAMP production (Rotrosen *et al.*, 1980). In the latter study, there was no effect of chronic ethanol exposure on β-adrenergic receptor-stimulated cAMP production in mesolimbic or cerebellar membranes. In contrast, however, chronic ethanol treatment decreased corticotropin-releasing hormone-dependent cAMP production by 40% in the neurointermediate lobe of the pituitary (Dave *et al.*, 1986). Reduced dopamine-stimulated cAMP production was also reported in striatal tissue of mice treated chronically with ethanol (Hoffman and Tabakoff, 1977; Tabakoff and Hoffman, 1979; Saffey *et al.*, 1988; Lucchi *et al.*, 1983). Since, in most cases, acute ethanol exposure caused an increase in receptor-stimulated cAMP production, the decrease in receptor-stimulated cAMP production observed with dopamine, as well as noradrenaline (Israel *et al.*, 1972; French *et al.*, 1975; Banerjee, *et al.*, 1978) and corticotropin-releasing hormone (Dave *et al.*, 1986), suggest a compensatory adaptive effect similar to agonist-induced heterologous desensitization (Sibley and Lefkowitz, 1985).

2.1.4 Receptors coupled to G_s

In rat cortical membranes, low ethanol concentrations decreased the affinity of the high-affinity form of the β-adrenergic receptor for agonists, but did not alter antagonist binding (Valverius *et al.*, 1987). These results suggest an effect of ethanol on the agonist–receptor–G_s complex rather than on receptor binding *per se*. Chronic exposure of mice to ethanol caused a decrease in high-affinity β-adrenergic receptor binding in hippocampus but no change in the number of receptors (Valverius *et al.*, 1987, 1989). In the cerebellum, however, there was a 16% decrease in the total number of β-adrenergic receptors (Valverius *et al.*, 1989). Inconsistent and variable results in different brain regions have also been reported in binding studies of the dopamine, serotonin, acetylcholine and γ-aminobutyric acid receptors (Muller *et al.*, 1980; Volicer, 1980; Yoshida *et al.*, 1982; Chung *et al.*, 1989).

2.1.5 G_s

The data reviewed above on the effects of ethanol on the cAMP signal transduction system suggest that G_s is a major target for ethanol. Ethanol could affect G_s activity by altering binding or hydrolysis of GTP or by altering the coupling of G_s to adenylyl cyclase. In mouse striatal membranes, acute ethanol did not directly affect the binding of Gpp(NH)p, a non-hydrolysable GTP analogue, to G_s in crude synaptosomal membranes from rat brain (Luthin and Tabakoff, 1984). However, high concentrations of ethanol did enhance the rate of G_s coupling to adenylyl cyclase in the presence of Gpp(NH)p (Luthin and Tabakoff, 1984). In membranes prepared from cerebral cortex, ethanol increased the efficacy of Gpp(NH)p stimulation of adenylyl cyclase (Saito *et al.*, 1985). Rabin and Molinoff (1983) have also concluded that G_s function is altered by ethanol, based on their studies of dopamine receptor-stimulated cAMP production in mouse striatal membranes. Using membranes from mouse cerebral cortex, Saito *et al.* (1985) proposed multiple sites of action to account for ethanol stimulation of adenylyl cyclase activity, including the rate of activation of G_s by Gpp(NH)p and G_s coupling to adenylyl cyclase. After chronic exposure to ethanol, Hoffman *et al.* found diminished ADP-ribosylation of G_s by cholera toxin in cortical membranes from ethanol-fed mice (Tabakoff *et al.*, 1988), which could be due either to a decrease in the amount of $G_s\alpha$ or an increase in endogenous ADP-ribosylation.

2.1.6 Summary

In summary, most studies in brain indicate that cAMP signal transduction, particularly the coupling of G_s to adenylyl cyclase, is a major target for ethanol. However, the variability of results in different brain regions and from different animal models preclude definitive statements on the mechanism underlying the effects of ethanol on the cAMP signal transduction system.

2.2 Cultured cell lines

Recent advances in understanding the interaction of ethanol with the cAMP signal transduction system derive from studies with intact cells in culture, including neural cell lines, S49 lymphoma cells and variants of S49 with altered cAMP signal transduction, and primary cerebellar and pineal cell cultures. Such studies have multiple advantages over experiments in intact animals: (1) the ethanol concentration is controlled and constant; (2) there is a homogeneous population of cells; (3) secondary effects due to ethanol such as alteration of hormone levels and trans-synaptic effects can be eliminated; (4) mutant cells can be used to determine the role of a specific gene product; and (5) intact cells can be used. It is important to use intact cells since low concentrations of ethanol alter cAMP signal transduction in intact cells and have no effect on isolated membrane preparations (Stenstrom and Richelson, 1982), suggesting that soluble cytoplasmic components are involved. Major advances in understanding the molecular basis of the effects of ethanol on the cAMP signal transduction system have been achieved using intact cells in culture.

2.2.1 Receptor-dependent cAMP production

Richelson and collaborators (Stenstrom and Richelson, 1982) were the first to use intact cultured neural cell lines to study the effects of ethanol on the cAMP signal transduction system. In NIE-115 cells, acute ethanol exposure increased PGE_1 receptor-stimulated cAMP production. We then showed that acute treatment with ethanol caused a 50% increase in adenosine receptor-stimulated cAMP levels in the NG108-15 neuroblastoma $\times$ glioma cell line (Gordon *et al.*, 1986). In contrast, chronic treatment with ethanol caused a 30% decrease in adenosine receptor-stimulated cAMP levels (desensitization) when measured in the absence of ethanol. The ethanol-induced desensitization was heterologous since PGE_1 receptor-dependent cAMP accumulation was also reduced to the same extent as adenosine receptor-stimulated cAMP levels in NG108-15 cells (Mochly-Rosen *et al.*, 1988). Acute stimulation of receptor-dependent cAMP levels was the same in control cells and those chronically exposed to ethanol (50% for both). After chronic exposure to ethanol, receptor-dependent cAMP levels were the same as in control cells only when the chronically treated cells were assayed in the presence of ethanol. Therefore, the chronic ethanol-induced decrease in cAMP signal transduction appeared to represent cellular dependence on alcohol. The effects of ethanol are reversible; receptor-dependent cAMP production returned to control levels within 48 h after removal of ethanol (Gordon *et al.*, 1986).

Many of the findings in NG108-15 cells have been confirmed, with some exceptions, by other investigators using the same (Charness *et al.*, 1988) or different cell lines (Richelson *et al.*, 1986; Charness *et al.*, 1988). Similar to the results in NG108-15 cells, Chung *et al.* (1989) have shown that 25 mM ethanol enhances agonist-stimulated cAMP levels in intact pineal glands in culture. Most importantly, ethanol-dependent increases in cAMP alter cellular function as evidenced from the finding that ethanol causes stimulation of cAMP-dependent melatonin production in these pineal cells.

In PC12 cells, a pheochromocytoma cell line with neuronal characteristics, several clones were differentially responsive to ethanol (Rabe, *et al.*, 1990). Receptor-stimulated cAMP levels in one subclone responded to acute and chronic ethanol treatment in a similar manner to that in NG108-15 cells (Rabe *et al.*, 1990; Rabin, 1988, 1990; Saito *et al.*, 1986). However, in a related subclone, acute ethanol exposure caused a decrease in adenosine receptor-stimulated cAMP production. Acute and chronic ethanol exposure alters receptor-dependent cAMP levels in primary cerebellar cultures, similar to its effects on NG108-15 cells (Rabin, 1990b). However, when primary cerebellar cultures exposed chronically to ethanol were rechallenged with ethanol, responses varied with different receptor agonists. Adenosine receptor-stimulated cAMP production was increased to the same extent on acute ethanol exposure in control and chronically treated cells as in NG108-15 cells, but β-adrenergic receptor stimulation after rechallenge with ethanol was further increased in cells that had been chronically exposed to ethanol.

2.2.2 *Adenylyl cyclcase*

In most studies, ethanol was found to have no effect on adenylyl cyclase activity *per se*. However, in membranes prepared from S49 *unc* and cyc⁻ cell mutants, where G_s is either uncoupled from receptor activation or absent, Rabin and Molinoff (1983) reported a dose-dependent increase in manganese-stimulated adenylyl cyclase activity upon acute treatment with ethanol. Since manganese uncouples G_s from adenylyl cyclase, this suggests that ethanol has a direct effect on the adenylyl cyclase catalytic subunit.

2.2.3 Receptors coupled to G_s

Acute treatment with ethanol of membranes prepared from S49 wild-type cells caused a decrease in the affinity of the β-adrenergic receptor for the agonist isoproterenol (Bode and Molinoff, 1988). A decrease in the affinity of the receptor for antagonists was also observed, but the magnitude of this effect was less than that for the agonists. Chronic ethanol treatment increased high affinity binding of agonist to the PGE_1 receptor in NIE-115 neuronal cells (Richelson *et al.*, 1986). However, no change in receptor density was observed, indicating that the expression and/or stability of receptors coupled to G_s is unaffected by chronic ethanol exposure.

2.2.4 G_s

Molinoff and co-workers introduced the use of S49 mouse lymphoma cell lines to address the possibility of direct effects of ethanol on G_s. Several cloned S49 cell lines with mutations in different components of the cAMP-dependent signalling system have been characterized (Johnson *et al.*, 1980). In *unc* cells, receptor activation is not coupled to adenylyl cyclase activation because of a mutation in α_s, the GTP-binding subunit of G_s (see Chapter 1). Using membranes prepared from wild-type and *unc* cells, Molinoff and co-workers (Rabin and Molinoff, 1983; Bode and Molinoff, 1988) demonstrated that acute ethanol increased G_s coupling to adenylyl cyclase. This was the first direct demonstration that G_s is the target for ethanol. Furthermore, G_s is also altered on chronic exposure to ethanol. We found that PGE_1 receptor-dependent cAMP production was decreased to the same extent as adenosine receptor-dependent cAMP production in NG108-15 cells (Mochly-Rosen *et al.*, 1988), suggesting that chronic ethanol caused a heterologous desensitization in these cells. These results suggest that chronic exposure to ethanol may have caused a decrease in $G_s\alpha$ amount and/or function which would compensate for the initial ethanol-induced increase in cAMP. We therefore measured $G_s\alpha$ mRNA and protein levels as well as $G_s\alpha$ function and found a 35% decrease in both $G_s\alpha$ mRNA and protein (Mochly-Rosen et al., 1988). To determine whether the decreased amount of protein had functional consequences, α_s was extracted from ethanol-treated and control membranes and its activity assayed by reconstitution with membranes from cyc⁻ cells that contain no $G_s\alpha$. We found 35% less adenylyl cyclase activity in cyc⁻ membranes reconstituted with α_s from ethanol-treated cells when compared to control cells. Since the decrease in receptor-stimulated cAMP production due to chronic ethanol exposure was of the same magnitude as the decrease in α_s function, it appears that the quantity of α_s is rate-limiting for cAMP production in these cells. Charness et al. (1988) confirmed the decrease in $G_s\alpha$ protein level after chronic ethanol exposure in NG108-15 cells and obtained similar results in N1E-115 cells. However, they observed no change in $G_s\alpha$ protein in N18TG2, the parental neuroblastoma cell of NG108-15.

2.2.5 Summary

In summary, a more coherent picture of the molecular events responsible for ethanol-induced changes in cAMP signal transduction emerges from studies with cultured cells. The most consistent finding is that acute ethanol exposure enhances coupling of $G_s\alpha$ to adenylyl cyclase and chronic ethanol exposure leads to a heterologous desensitization of receptors coupled to adenylyl cyclase which appears to be due to a decrease in α_s mRNA, protein, and function.

2.3 Human cells

Changes in the cAMP signal transduction system observed in brain homogenates and cultured cell lines due to ethanol exposure appear to be relevant to the pathophysiology of alcoholism. Clinical evidence indicates that alcoholics develop significant functional and structural abnormalities in most organs and tissues (Kissin, 1988). Although alcohol-induced changes in the brain account for intoxication and subsequent addiction to ethanol, the CNS is not accessible for direct study. Therefore, investigators have examined circulating blood cells, with the expectation that ethanol-induced changes in the cAMP signal transduction system of pathophysiological significance in brain would also be demonstrable in peripheral cells. Similar to results in neural cell lines, Atkinson *et al.* (1977) found that ethanol causes an increase in cAMP levels in human peripheral blood lymphocytes and Hynie *et al.* (1980) reported that ethanol potentiates adenosine and PGE_1 receptor-dependent cAMP production in the same cells. Moreover, ethanol stimulates adenylyl cyclase activity in human lymphocytes, apparently by facilitating nucleotide–G-protein interaction in lymphocyte membranes (Saito *et al.*, 1987). Since prolonged exposure of neural cells to ethanol causes heterologous desensitization of receptors coupled to adenylyl cyclase via $G_s\alpha$ (Mochly-Rosen *et al.*, 1988), we studied freshly isolated human peripheral blood lymphocytes from alcoholic subjects and non-alcoholic controls to determine whether chronic exposure to ethanol *in vivo* also desensitizes cAMP signal transduction (Diamond *et al.*, 1987). We found that there was a heterologous desensitization in intact lymphocytes from alcoholics; a 76% reduction in basal, adenosine and β-adrenergic receptor-stimulated cAMP production was observed in lymphocytes from actively drinking alcoholics when compared to lymphocytes from non-alcoholics or patients with non-alcohol-related liver disease (Figure 1). In contrast to cultured cells, lymphocytes from alcoholics were also 'tolerant' when rechallenged with ethanol; ethanol had little effect on adenosine receptor-dependent cAMP production (Figure 1). In a preliminary study, we found a 60% decrease in $G_s\alpha$ in membranes from freshly isolated lymphocytes of actively drinking alcoholics as compared to non-alcoholics (Figure 2). The correlation between reduced receptor-dependent cAMP levels and reduced $G_s\alpha$ protein levels in lymphocytes from alcoholics is similar to results obtained with NG108-15 cells exposed to ethanol for 48 h (Mochly-Rosen *et al.*, 1988). The increased magnitude of ethanol-induced desensitization and decrease in $G_s\alpha$ levels in lymphocytes from alcoholics relative to cultured cell lines was probably due to the consumption of large amounts of alcohol over many years by the alcoholics compared to a 48 h exposure to ethanol for the cultured cells. Consistent with our results with circulating lymphocytes, subsequent studies upon platelet membranes from alcoholics showed a lesser but significant reduction in PGE_1, guanine nucleotide, and fluoride-stimulated adenylyl cyclase activity (Tabakoff *et al.*, 1988b).

Heterologous desensitization of cAMP production in alcoholics may be a consequence of heavy drinking. It is also possible that the same membrane-dependent events which are affected by ethanol are more sensitive to ethanol in people at risk of developing alcoholism due to genetic alterations. This question cannot be studied in actively drink-

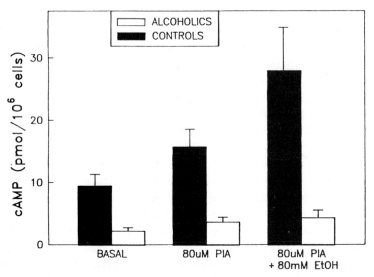

Figure 1. Basal and phenylisopropyladenosine (PIA)-stimulated cAMP levels in lymphocytes from alcoholics (open bars) and control subjects (solid bars). PIA is an adenosine receptor agonist. Each bar represents the mean $\pm$ SEM ($n = 10$ for basal and PIA; $n = 9$ for PIA plus ethanol).

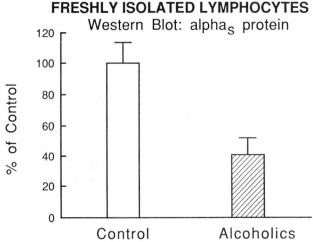

Figure 2. $G_s\alpha$ protein in lymphocytes from alcoholics (open bars) and controls (striped bars). Membranes were prepared from lymphocytes of seven alcoholics and nine control subjects. After SDS polyacrylamide gel electrophoresis, $G_s\alpha$ levels were determined by Western blot analysis using anti-$G_s\alpha$ antibodies. Values reported are mean $\pm$ SEM, relative to those of controls.

ing alcoholics because of their concomitant exposure to alcohol, and the metabolic, hormonal and nutritional complications encountered in alcoholism. In order to distinguish between a 'toxic' effect of ethanol due to *in vivo* exposure and a possible genetic difference in alcoholics, we cultured lymphocytes from alcoholics and non-alcoholics for 7–8 days without ethanol and then measured receptor-dependent cAMP levels (Nagy *et al.*, 1988). After four to six cell divisions in culture without ethanol, adenosine receptor-stimulated cAMP production in lymphocytes from alcoholics was 2.8-fold higher than in cells from non-alcoholics. This increase in cultured cells grown in the absence of ethanol was specific for the adenosine receptor. There was no difference in PGE_1 receptor-dependent cAMP accumulation in cells from alcoholics and non-alcoholics (L. E. Nagy, I. Diamond and A. S. Gordon, unpublished observations). Therefore, decreased cAMP production in circulating cells from alcoholics appears to be due to *in vivo* ethanol exposure. It is possible that measurement of decreased receptor-stimulated cAMP levels could serve as a potential biological marker for active alcohol consumption.

To determine whether alcoholics have increased sensitivity to ethanol, lymphocytes were grown in 100 mM ethanol for 24 h (Nagy *et al.*, 1988). Under these conditions, there was no change in adenosine receptor-dependent cAMP levels in cells from non-alcoholics. However, lymphocytes from alcoholics showed a desensitization of cAMP production. There was a 39% reduction in adenosine receptor-stimulated cAMP levels (Figure 3). Taken together, the results show that despite four to six cell divisions in culture without ethanol, where more than 94% of the cells had never been exposed to ethanol, lymphocytes from alcoholics exhibit significantly increased adenosine receptor-dependent cAMP levels and increased sensitivity when exposed chronically to ethanol. These findings suggest that the regulation of cAMP signal transduction may be altered in patients at risk of developing alcoholism. Moreover, these differences in cultured cells from alcoholics appear to constitute a phenotypic abnormality which may identify patients vulnerable to alcoholism because of genetic factors.

2.4 The role of adenosine in mediating ethanol-induced changes in cAMP signal transduction

Acute ethanol exposure causes an increase in receptor-dependent cAMP accumulation. After chronic exposure to ethanol, there is a heterologous desensitization of cAMP accumulation on activation of receptors coupled to $G_s\alpha$. By analogy with other models of desensitization, we proposed that the chronic effects of ethanol are due to a compensatory cellular response to the acute increase in cAMP. To determine the molecular basis of the acute and adaptive responses to ethanol, we have used NG108-15 cells and the S49 cell mutants.

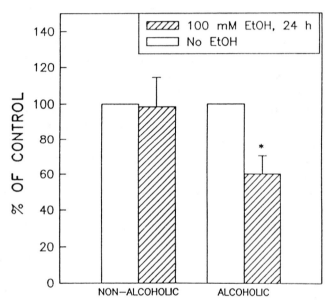

Figure 3. Difference in sensitivity to chronic ethanol exposure between cells from alcoholic and non-alcoholic subjects. Lymphocytes from alcoholic ($n = 7$) and non-alcoholic ($n = 9$) subjects were grown in culture for 7 days; 100 mM ethanol was added to half of the culture flasks for the last 24 h, and PIA-stimulated cAMP levels were determined. Values represent cAMP levels in ethanol-treated cells as a percentage of cAMP levels in cells grown without ethanol. * $p < 0.003$ (compared with non-alcoholic subjects).

2.4.1 Cultured cell lines

Acute ethanol exposure Heterologous desensitization in many cell types is characterized by an initial increase in cAMP levels (Sibley and Lefkowitz, 1985). Therefore, we determined whether acute exposure to ethanol caused an increase in cellular cAMP levels in intact cells in the absence of exogenously added agonist. When NG108-15 cells were incubated with 100 mM ethanol for 10 min, there was a 60% increase in intracellular cAMP levels. Because ethanol does not activate adenylyl cyclase directly in these cells but stimulates only receptor-dependent cAMP production (Gordon *et al.*, 1986), we next explored the possibility that acute ethanol increases the extracellular concentration of a stimulatory agonist (Nagy *et al.*, 1989). Neural cells (Proctor and Dunwiddie, 1984), lymphocytes (Newby and Holmquist, 1981; Fredholm *et al.*, 1978) and other cell types (Stone, 1985) release adenosine, which can cause both homologous and heterologous desensitization (Kenimer and Nirenberg, 1981; Newman and Levitzki, 1983). Moreover, adenosine has been implicated in the effects of ethanol on the CNS (Proctor and Dunwiddie, 1984; Dar *et al.*, 1983). Therefore, adenosine concentrations in the media of control and ethanol-treated cells were measured using high pressure liquid chromatography. There was a significant increase in the concentration of extracellular adenosine after NG108-15 cells were incubated with 200 mM ethanol. When 5×10^6 cells were incubated in ethanol for 10 min, adenosine concentrations reached 37 ± 1.2 pmoles in 2 ml of media. In control cells, extracellular adenosine was 18.2 ± 3.7 pmoles (Figure 4).

In the NG108-15 cell line, acute ethanol exposure increases intracellular cAMP levels in the absence of added agonist. We found that this increase is due to an ethanol-dependent increase in the extracellular accumulation of adenosine in NG108-15 cells (Nagy *et al.*, 1989). If adenosine is degraded to inosine by adenosine deaminase or an adenosine receptor antagonist is present during the time of incubation with ethanol, no ethanol-dependent increase in cAMP is observed (Nagy *et al.*, 1989). These results suggest that acute ethanol exposure increases cAMP levels by causing an accumulation of extracellular adenosine. This adenosine then activates adenosine A_2 receptors coupled to adenylyl cyclase via G_s.

Since adenosine is continually being released from and taken up by most cell types via a nucleoside transporter, the initial target of ethanol action could be the nucleoside transporter. Figure 5 demonstrates that ethanol specifically inhibits uptake of adenosine via the nucleoside transporter. We also found the same inhibition of adenosine uptake by ethanol in NG108-15 cells and the human lymphoid cell line, CEM (Nagy *et al.*, 1990). In S49 cells, the inhibition of adenosine uptake by ethanol completely accounts for the accumulation of extracellular adenosine (Figure 4) and the ethanol-induced increases in cAMP levels.

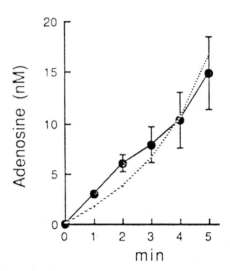

Figure 4. Time-dependent increase in extracellular adenosine concentration in the presence of 200 mM ethanol. Aliquots of S49 cells in suspension (2×10^7 cells/ml) were incubated without ethanol for 5 min. Ethanol was then added to a final concentration of 200 mM at 5, 6, 7, 8 or 9 min. Cells were removed from the ethanol-containing medium after a total incubation time of 10 min. Control cells were carried through the entire 10 min incubation without ethanol. Extracellular adenosine concentrations were measured by HPLC. The vertical axis represents the increase in extracellular adenosine in the supernatant of cells treated with ethanol above that found in the absence of ethanol (4.4 ± 0.9 nM). The dotted line represents the predicted increase in extracellular adenosine which would result from a 35% reduction in adenosine uptake. This line was calculated by increasing the 4.4 nM adenosine observed in the absence of ethanol by 35% over each minute of the incubation in ethanol. Values represent mean $\pm$ SEM, $n = 5$.

We have recently found that cAMP-dependent protein kinase activity is required for ethanol to inhibit adenosine uptake (L.N. Nagy, I. Diamond and A.S. Gordon, unpublished observations). In mutant S49 cell lines lacking cAMP-dependent protein kinase activity, ethanol does not inhibit adenosine uptake nor increase extracellular adenosine accumulation. In *unc* cells, an S49 variant in which α_s is present, but uncoupled to receptor stimulation, ethanol also does not inhibit adenosine uptake. However, in the presence of forskolin, which bypasses the α_s defect to increase cAMP concentration, inhibition of adenosine uptake by ethanol is restored. We have proposed that phosphorylation of either the adenosine transporter or a regulatory

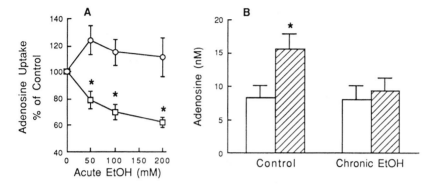

Figure 5. Acute and chronic effects of ethanol on adenosine uptake and extracellular adenosine concentration. (A) Chronic effects of ethanol on [³H]adenosine uptake. S49 cells were grown in the absence (squares) or presence (circles) of 100 mM ethanol for 48 h and adenosine uptake determined. Uptake was measured at 90 s in the presence of 50–200 mM ethanol and compared to cells incubated in the absence of ethanol. Values represent mean ± SEM, $n = 6$–11. *$p < 0.01$ compared to uptake in the absence of ethanol. (B) Ethanol-induced extracellular adenosine accumulation in S49 cells. Cells were grown in the absence (control) or presence (chronic EtOH) of 100 mM ethanol for 48 h. Media were removed and S49 cells (1×10^7 cells/ml) incubated in the absence (open bars) or presence (striped bars) of 100 mM ethanol for 10 min. Values represent mean ± SEM, $n = 10$–14. *$p < 0.003$ compared to cells not treated with acute ethanol.

component is required for inhibition of adenosine uptake on acute exposure to ethanol.

Chronic ethanol exposure Chronic exposure to ethanol causes a heterologous desensitization of receptors coupled to adenylyl cyclase via G_s. We have shown that concomitant with this desensitization, there is a decrease in α_s mRNA and protein as well as α_s function when measured in a reconstitution assay. In addition, we have recently found that the adenosine transporter becomes insensitive to rechallenge with ethanol after chronic exposure to ethanol; adenosine uptake is no longer inhibited by ethanol and no accumulation of extracellular adenosine occurs (Nagy *et al.*, 1990) (Figure 5). Similar to the acute effects of ethanol in the cAMP signal transduction system, the chronic effects appear to be mediated by adenosine. Heterologous desensitization does not occur if adenosine deaminase (Nagy *et al.*, 1989) or adenosine receptor antagonists (M. Sapru, I. Diamond and A.S. Gordon, unpublished results) are present during the ethanol incubation. Moreover, when a mutant S49 cell line lacking the nucleoside transporter is incubated with ethanol for 48 h, no heterologous desensitization occurs (Nagy *et al.*, 1989). Co-incubation of NG108-15 cells with ethanol and an adenosine receptor antagonist prevents not only heterologous desensitization, but also the decrease in the amount of α_s. Moreover, when cells are co-incubated with ethanol and an adenosine receptor antagonist for 48 h, no loss of sensitivity of adenosine transport to ethanol occurs and extracellular adenosine still accumulates on acute exposure to ethanol (M. Sapru, I. Diamond and A.S. Gordon, unpublished observations).

In PC12 cells (Rabe *et al.*, 1990) and primary cerebellar neurones from rat (Rabin, 1990), heterologous desensitization does not appear to be due to the action of extracellular adenosine on adenosine A_2 receptors. However, since both these cell types release various neurotransmitters and/or neuropeptides, and neurotransmitter release has been shown to be increased in the presence of ethanol (Lynch *et al.*, 1986; Rabe and Weight, 1988), changes in intracellular cAMP could occur as a result of interaction of the released compounds with any corresponding receptors on these cells. The relative effects of adenosine and other neuropeptides and neurotransmitters which accumulate extracellularly on altered cAMP levels which occur after ethanol exposure would depend on the specific ratio of accumulated agonists and their respective receptors present in the vicinity of their release.

Taken together, our results suggest that interaction of ethanol with the nucleoside transporter is the initial step in the cascade of events caused by ethanol (Figure 6). Inhibition of adenosine influx results in an accumulation of extracellular adenosine, which in turn, activates adenosine A_2 receptors to acutely increase intracellular cAMP levels. Ethanol further increases cAMP accumulation by enhancing coupling of $G_s\alpha$ to adenylyl cyclase. The chronic effects of ethanol appear to be due to adenosine accumulation, since blocking the acute effects of ethanol at the level of either the adenosine transporter or the adenosine receptor also prevents the chronic effects. We propose that, similar to other systems where heterologous desensitization occurs (Sibley and Lefkowitz, 1985), incubation with ethanol leads to activation of cAMP-dependent protein kinase due to ethanol-induced inhibition of adenosine transport and subsequent increase in intracellular cAMP levels. Increased protein kinase A (PKA) activity leads to an adaptive response, whereby α_s mRNA, protein and function are decreased, resulting in decreased receptor-stimulated cAMP levels.

Concomitantly, and presumably due to the decreased cAMP level, the adenosine transporter becomes insensitive to rechallenge with ethanol and there is no longer an increase in extracellular adenosine concentration.

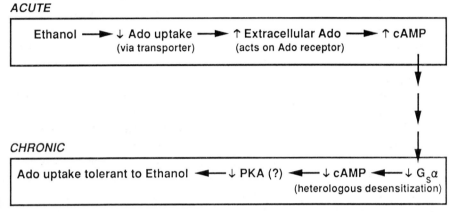

Figure 6 Model for the role of adenosine (Ado) in mediating the acute and chronic effects of ethanol on the cAMP signal transduction system.

2.4.2 Human cells

Recent evidence suggests that adenosine, an inhibitory neuromodulator, mediates ethanol-induced motor disturbances (Dar *et al.*, 1983; Dar and Wooles, 1986) and sedation in mice (Proctor and Dunwiddie, 1984). Dar and colleagues have found that pretreatment of rats with theophylline, an adenosine receptor antagonist, decreases acute ethanol-induced incoordination (Dar *et al.*, 1983; Dar and Wooles, 1986). In contrast, dipyridamole, which inhibits adenosine uptake and potentiates the effects of adenosine, exacerbates the acute effects of ethanol on motor function (Dar *et al.*, 1983; Dar and Wooles, 1986), suggesting that extracellular adenosine may be involved in the neurological response to alcohol.

Since both the acute and chronic effects of ethanol on cellular cAMP levels in cultured cell lines are mediated by adenosine, the differences in cAMP signal transduction in cells from alcoholics might be due to altered sensitivity of adenosine transport to ethanol. Recently, we have identified differences in adenosine transport in freshly isolated lymphocytes from alcoholics as compared to non-alcoholics (L.E. Nagy, S.K. Krauss, I. Diamond and A.S. Gordon, unpublished data). Adenosine uptake in lymphocytes from non-alcoholics is inhibited by 70% on acute exposure to ethanol. In contrast, adenosine uptake in lymphocytes from alcoholics was not inhibited by ethanol, i.e. was tolerant to ethanol. Moreover, adenosine uptake in lymphocytes from alcoholics when measured in the absence of ethanol was reduced by 41% in comparison to cells from non-alcoholics. To determine whether these differences are due to *in vivo* ethanol exposure or a possible genetic difference, we examined adenosine uptake in cultured lymphocytes. After growth in the absence of ethanol for four to six cell divisions, there was no longer any difference in adenosine uptake in the absence of ethanol between lymphocytes from alcoholics and non-alcoholics, indicating that reduced uptake of adenosine in circulating lymphocytes from alcoholics is probably due to *in vivo* ethanol exposure (see Section 2.3). As in circulating lymphocytes, acute ethanol exposure of cultured cells from non-alcoholics *in vitro* resulted in inhibition of adenosine uptake. However, adenosine uptake in cultured lymphocytes from alcoholics was still not inhibited by ethanol.

Thus, cultured lymphocytes from alcoholics appear to be tolerant to ethanol with respect to adenosine uptake. These results suggest that differences in regulation of adenosine transport by ethanol in cells from alcoholics may be due to genetic alterations. Such information may lead to the identification of a biological marker for a genetic predisposition to alcoholism as well as new therapeutic approaches to the treatment of alcoholism and alcohol dependency.

3 SUMMARY

Acute exposure to ethanol in cultured cell lines inhibits adenosine uptake into cells, thereby increasing the concentration of extracellular adenosine. Extracellular adenosine then binds to adenosine A_2 receptors to stimulate intracellular cAMP production. During prolonged exposure to ethanol, the increase in cAMP is followed by the development of heterologous desensitization of receptors coupled to adenylyl cyclase via G_s. Heterologous desensitization appears to be due to a reduction in mRNA and protein for $G_s\alpha$. This is an example of cellular dependence on ethanol since cells are 'normal' only in the presence of ethanol. The important implication of these findings is that a selective inhibitory effect of ethanol on adenosine uptake can lead to widespread desensitization of receptors coupled to cAMP production. Such changes could contribute to the pleiotropic effects of ethanol in the brain and other organs.

Prolonged exposure to ethanol also alters the nucleoside transport system in cultured cell lines. Whereas ethanol inhibits adenosine uptake into naive cells, ethanol no longer inhibits adenosine uptake into cells that have adapted to ethanol. This resistance to ethanol inhibition appears to be a form of cellular tolerance to ethanol. As a result, extracellular adenosine levels are no longer increased by ethanol, thereby reducing adenosine receptor stimulation of adenylyl cyclase.

Studies on cAMP signal transduction in cell culture are directly relevant to the pathophysiology of human alcoholism. Heterologous desensitization of cAMP production is demonstrable in lymphocytes taken from actively drinking alcoholics; this measurement may be a useful biological marker for active alcohol consumption. In addition, regulation of adenosine receptor-dependent cAMP production may be altered in patients at risk of developing alcoholism due to a genetic predisposition. We found that lymphocytes from alcoholics cultured for many generations in the absence of ethanol show increased adeno-sine receptor-stimulated cAMP production and increased sensitivity to ethanol-induced heterologous desensitization. In addition, cultured cells from alcoholics grown in the absence of ethanol are insensitive to the inhibitory effects of ethanol on adenosine uptake. Therefore, these persistent phenotypic abnormalities in cell culture could be genetic markers for a predisposition to alcoholism.

REFERENCES

Atkinson, J.P., Sullivan, T.J., Kelly, J.P. and Parker, C.W. (1977). *J. Clin. Invest.* **60**, 284–294.

Banerjee, S.P., Sharma, V.K. and Khanna, J.M. (1978). *Nature* **276**, 407–409.

Bode, D.C. and Molinoff, P.B. (1988). *J. Pharmacol. Exp. Ther.* **246**, 1040–1047.

Charness, M.E., Querimit, L.A. and Diamond, I. (1986). *J. Biol. Chem.* **261**, 3164–3169.

Charness, M.E., Querimit, L.A. and Henteleff, M. (1988). *Biochem. Biophys. Res. Commun.* **155**, 138–143.

Chung, C.T., Tamarkin, L., Hoffman, P.L. and Tabakoff, B. (1989). *J. Pharma-col. Exp. Ther.* **249**, 16–22.

Cloninger, C.R. (1987). *Science* **236**, 410–416.

Daniell, L.C. and Leslie, S.W. (1986). *Brain Res.* **337**, 18–28.

Dar, M.S. and Wooles, W.R. (1986). *Life Sci.* **39**, 1429–1437.

Dar, M.S., Mustafa, S.J. and Wooles, W.R. (1983). *Life Sci.* **33**, 1363–1374.

Dave, J.R., Eiden, L.E., Karanian, J.W. and Eskay, R.L. (1986). *Endocrinology* **118**, 280–286.

Devor, E.J. and Cloninger, C.R. (1989). *Annu. Rev. Genet.* **23**, 19–35.

Diamond, I., Wrubel, B. and Estrin, E. (1987). *Proc. Natl. Acad. Sci. USA* **84**, 1413–1416.

Farrar, R.P., Seibert, C. and Gnau, K. (1989). *Brain Res.* **500**, 374–378.

Ferko, A.P., Bobyock, E. and Chernick, W.S. (1982). *Toxicol. Appl. Pharmacol.* **64**, 447–455.

Fredholm, B.B., Sandberg, G. and Ernstrom, U. (1978). *Biochem. Pharmacol.* **27**, 2675–2682.

French, S.W., Palmer, D.S., Narod, M.E., Reid, P.E. and Ramey, C.W. (1975). *J. Pharmacol. Exp. Ther.* **194**, 319–326.

Goodwin, D.W. (1987). *Adv. Intern. Med.* **32**, 283–298.

Goldstein, D. B. (1983). *Pharmacology of Alcoholism.* Oxford University Press, New York.

Goldstein, D.B. (1986). *Ann. N. Y. Acad. Sci.* **492**, 103–111.

Goldstein, D.B. and Chin, J.H. (1981). *Fed. Proc.* **40**, 2073–2076.

Gordon, A.S., Collier, K. and Diamond, I. (1986). *Proc. Natl. Acad. Sci. USA* **83**, 2105–2108.

Hoffman, P.L. and Tabakoff, B. (1977). *Nature* **268**, 551–553.

Hoffman, P.L. and Tabakoff, B. (1986). *J. Neurochem.* **46**, 812–816.

Hoffman, P.L., Luthin, G. and Tabakoff, B. (1982). *Alcoholism: Clin. Exp. Res.* **6**, 300–305.

Hoffman, P.L., Saito, T. and Tabakoff, B. (1987). *Ann. N. Y. Acad. Sci.* **492**, 396–397.

Hynie, S., Lanefelt, F. and Fredholm, B.B. (1980). *Acta Pharmacol. Toxicol.* **47**, 58–65.

Israel, M.A., Kimura, H. and Kuriyama, K. (1972). *Experientia* **28**, 1322–1323.

Johnson, G.L., Kaslow, H.R., Farfel, Z. and Bourne, H.R. (1980). *Adv. Cyclic Nucleotide Res.* **13**, 1–37.

Johnson, R.A., Noll, E.C. and Rodney, W.M. (1982). *Lancet* **ii**, 1394.

Kenimer, J.G. and Nirenberg, M. (1981). *Mol. Pharmacol.* **20**, 585–591.

Kiianmaa, M., Tabakoff, B. and Saito, T. (eds) (1989). *Genetic Aspects of Alcoholism.* Finnish Foundation for Alcohol Studies, Helsinki.

Kissin, B. (1988). In: *Alcohol Abuse and Alcohol-related Illnesses* (Wyngaarden, J.B. and Smith, L.H., eds), pp. 48–52. W. B. Saunders, Philadelphia.

Linblad, B. and Olsson, R. (1976). *J. Am. Med. Assoc.* **236**, 1600–1602.

Lucchi, L., Covelli, V., Anthopoulou, H., Spano, P.F. and Trabucchi, M. (1983). *Neurosci. Lett.* **40**, 187–192.

Luthin, G.R. and Tabakoff, B. (1984). *J. Pharmacol. Exp. Ther.* **228**, 579–587.

Lynch, M.A., Archer, E.R. and Littleton, J.M. (1986). *Biochem. Pharmacol.* **35**, 1207–1209.

Messing, R.O., Carpenter, C.L., Diamond, I. and Greenberg, D. A. (1986). *Proc. Natl. Acad. Sci. USA* **83**, 6213–6215.

Mhatre, M. and Ticku, M.K. (1989). *J. Pharmacol. Exp. Ther.* **251**, 164–168.

Miles, M.F. and Sturdivant, J. (1991). *J. Biol. Chem.* **266**, 2409–2414.

Mirsky, I.A., Piker, P. and Rosenbaum. M. (1941). *Q. J. Stud. Alcohol* **2**, 35–45.

Mochly-Rosen, D., Chang, F.-H., Cheever, M., Kim, I., Diamond, I. and Gordon, A.S. (1988). *Nature* **333**, 848–850.

Muller, P., Britton, R.S. and Seeman, P. (1980). *Eur. J. Pharmacol.* **65**, 31–37.

Nagy, L.E., Diamond, I. and Gordon, A.S. (1988). *Proc. Natl. Acad. Sci. USA* **85**, 6973–6976.

Nagy, L.E., Diamond, I., Collier, K., Lopez, L., Ullman, B. and Gordon, A.S. (1989). *Mol. Pharmacol.* **36**, 744–748.

Nagy, L.E., Diamond, I., Casso, D.J., Franklin, C. and Gordon, A.S. (1990). *J. Biol. Chem.* **265**, 1946–1951.

Newby, A.C. and Holmquist, C.A. (1981). *Biochem. J.* **200**, 399–403.

Newman, M.E. and Levitzki, A. (1983). *Biochem. Pharmacol.* **32**, 137–140.

Okuda, C., Miyazaki, M. and Kuriyama, K. (1984). *Neurochem. Int.* **6**, 237–244.

Porter, R., Matson, R. and Kramer, J. (eds) (1991). *Alcohol and Seizures: Basic Mechanisms and Clinical Concepts.* F. A. Davis, Philadelphia.

Proctor, W.R. and Dunwiddie, T.V. (1984). *Science* **224**, 519–521.

Rabe, C.S. and Weight, F.F. (1988). *J. Pharmacol. Exp. Ther.* **244**, 417–422.

Rabe, C.S., Giri, P.R., Hoffman, P.L. and Tabakoff, B. (1990). *Biochem. Pharmacol.* **40**, 565–571.

Rabin, R.A. (1985). *Biochem. Pharmacol.* **34**, 4329–4331.

Rabin, R. A. (1988). *J. Neurochem.* **51**, 1148–1155.

Rabin, R. A. (1990a). *J. Pharmacol. Exper. Ther.* **252**, 1021–1027.

Rabin, R. A. (1990b). *J. Neurochem.* **55**, 122–128.

Rabin, R.A. and Molinoff, P.B. (1981). *J. Pharmacol. Exp. Ther.* **216**, 129–134

Rabin, R.A. and Molinoff, P.B. (1983). *J. Pharmacol. Exper. Ther.* **227**, 551–556.

Rabin, R.A., Bode, D.C. and Molinoff, P.B. (1986). *Biochem. Pharmacol.* **35**, 2331–2335.

Reynolds, J.N., Wu, P.H. and Khanna, J.M. (1990). *J. Pharmacol. Exp. Ther.* **252**, 265–271.

Richelson, E., Stenstrom, S., Forray, C., Enloe, L. and Pfenning, M. (1986). *J. Pharmacol. Exp. Ther.* **239**, 687–692.

Rius, R.A., Bergamaschi, S., Lucchi, L., Govoni, S. and Trabucchi M. (1987). *Alcohol Alcoholism* Suppl. 1, 755–759.

Rotrosen, J., Mandio, D., Segarnick, D., Traficante, L.J. and Gershon, S. (1980). *Life Sci.* **26**, 1867–1876.

Rottenberg, H. (1986). *Biochim. Biophys. Acta* **855**, 211–222.

Saffey, K., Gillman, M. A. and Cantrill, R.C. (1988). *Neurosci. Lett.* **84**, 317–322.

Saito, T., Lee, J.M. and Tabakoff, B. (1985). *J. Neurochem.* **44**, 1037–1044.

Saito, T., Tsuchiya, F., Ishizawa, H. and Hatta, Y. (1986). *Jpn. J. Alcohol Drug Dependence* **21**, 135–145.

Saito, T., Lee, J.M., Hoffman, P.L. and Tabakoff, B. (1987). *J. Neurosci.* **48**, 1817–1822.

Shen, A., Jacobyansky, A., Pathman, D. and Thurman, R.G. (1983). *Eur. J. Pharmacol.* **89**, 103–110.

Sibley, D.R. and Lefkowitz, R.J. (1985). *Nature* **317**, 124–129.

Smith, T.L. (1981). *Proc. West. Pharmacol. Soc.* **24**, 37–39.

Smith, T.L., Jacobyansky, A., Shen., A., Pathman, D. and Thurman, R.G. (1981). *Neuropharmacology* **20**, 67–72.

Stenstrom, S. and Richelson, E. (1982). *J. Pharmacol. Exp. Ther.* **221**, 334–341.

Stenstrom, S., Enloe, L. and Pfenning, M. (1986). *J. Pharmacol. Exp. Ther.* **236**, 458–463.

Stone, T.W. (ed.) (1985). *Purines: Pharmacological and Physiological Roles.* Macmillan, London.

Sun, G.Y. and Sun, A.Y. (1985). *Alcoholism: Clin. Exp. Res.* **9**, 164–180.

Tabakoff, B. and Hoffman, P.L. (1979). *J. Pharmacol. Exp. Ther.* **208**, 216–222.

Tabakoff, B. and Hoffman, P.L. (1983). *Life Sci.* **32**, 197–204.

Tabakoff, B.P., Valverius, P., Nhamburo, P.T., Rius, R.A. and Hoffman, P.L. (1988). *Trans. Am. Soc. Neurochem.* **19**, 144.

Tabakoff, B., Hoffman, P.L., Lee, J.M., Saito, T., Willard, B. and De Leon-Jones, F. (1988). *N. Engl. J. Med.* **318**, 134–139.

Taraschi, T.F., Ellingson, J.S. and Wu, A. (1986). *Proc. Natl. Acad. Sci. USA* **83**, 9398–9402.

Urso, T., Gavaler, J.S. and Van Thiel, D.H. (1981). *Life Sci.* **28**, 1053–1056.

Valverius, P., Hoffman, P.L. and Tabakoff, B. (1987). *Mol. Pharmacol.* **32**, 217–222.

Valverius, P., Hoffman, P.L. and Tabakoff, B. (1989). *J. Neurochem.* **52**, 492–497.

Victor, M. and Adams, R.D. (1953). *Assoc. Res. Nerv. Mental Disorders* **32**, 526–573.

Volicer, L. (1980). *Brain Res. Bull.* **5**, 809–813.

Watanabe, A., Kobayashi, M. and Hobara, N. (1985). *Alcoholism: Clin. Exp. Res.* **9**, 14–16.

White, G., Lovinger, D.M. and Weight, F.F. (1990). *Brain Res.* **507**, 332–336.

Yoshida, K., Engel, J. and Liljequist, S. (1982). *Naunyn-Schmiedebergs Arch. Pharmacol.* **321**, 74–76.

CHAPTER NINE

The Role of G-proteins in the Regulation of Myeloid Haemopoietic Cell Proliferation and Transformation

PHILIP MUSK, CLARE M. HEYWORTH,*
SUSAN J. VALLANCE and ANTHONY D. WHETTON
*Department of Biochemistry and Applied Molecular
Biology, U.M.I.S.T., Sackville Street, Manchester,
M60 1QD, UK*
* *Experimental Haematology Department, Paterson
Institute, Christie Hospital and Holt Radium Insti-
tute, Manchester, M20 9BX, UK*

1 HAEMOPOIESIS

Haemopoiesis is the process of blood cell production. Recently, a number of growth factors which regulate the development of haemopoietic cells *in vivo* and *in vitro* have been isolated and cloned. The molecular mechanisms whereby these growth factors promote such events are at present unclear, but there is evidence that guanyl nucleotide regulatory (G) proteins play some role in these processes. There is also evidence that the discordant development observed in myeloid leukaemias is associated with the inappropriate activation of one particular G-protein, *ras*. In this review, the regulation of myeloid blood cell production will be described and the relevance of guanyl nucleotide regulatory proteins to the development of blood cells and leukaemia examined.

1.1 The multipotent stem cell

One of the early arguments concerning the haemopoietic system was on the origin of blood cells. Was there a common ancestral cell for all

G-Proteins
ISBN 0-12-497515-1

the blood cell types, or did the bone marrow contain a specific precursor cell for each type of peripheral blood cell? The most important piece of evidence which led to confirmation of the idea that pluripotent haemopoietic cells existed in adult mice was the establishment of the spleen colony-forming assay by McCulloch and Till (1964). When irradiated mice (in which the haemopoietic system was fatally damaged) were injected with fresh normal bone marrow cells, nodules were seen to develop on the spleen. These nodules contained mature cells from several distinct lineages and also cells which were capable of generating further spleen colonies, termed colony-forming unit – spleen (CFU-S). The evidence derived from experiments employing this assay indicated that there is a population of multipotent cells present within the marrow which can self-renew, effectively proving the existence of the common ancestral cell within the bone marrow. The CFU-S assay also provided data on the kinetics of the multipotent cells. In the healthy mouse these cells were generally not in cycle, residing mainly in Go. In experiments where high doses of [^{3}H]-thymidine were applied *in vivo* or *in vitro*, less than 10% of the CFU-S were killed (i.e. were actively proliferating). The application of cytotoxic drugs or radiation to mice can increase the number of CFU-S in cycle by five times, indicating that this population can be triggered into cycle in order to redress any imbalance caused by injury or insult to the haemopoietic system (Lajtha, 1982). Similarly, lethally irradiated mice engrafted with normal bone marrow show a steady increase in the CFU-S population (doubling about every 24 hours) until a new steady state is realized, at which point these cells revert to a predominantly non-cycling population. It has been suggested that some agonists which are coupled to G-proteins (e.g. β-adrenergic agents and histamine; see Byron (1975, 1977)) can induce CFU-S cycling, but it is improbable that this is achieved via a direct effect on these cells. There are, however, a number of growth factors and cytokines which can influence the proliferation, differentiation and development of these multipotent cells (see below).

The low level of CFU-S that are in cycle at any time, coupled with the observation that there are only about 5000 CFU-S per 20 million bone marrow cells in mice, suggests that the large numbers of developing (e.g. erythroblasts, promyelocytes) and mature haemopoietic cells found in the bone marrow and peripheral blood must be formed via the proliferation of transitional populations of developing blood cells, which amplify the number of CFU-S present (see Figure 1).

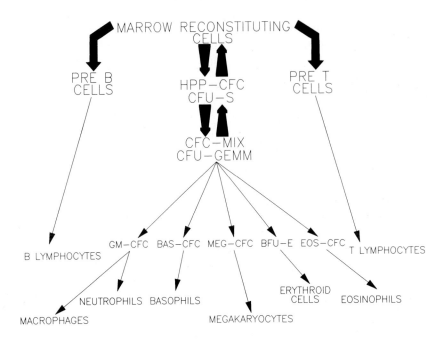

Figure 1. The structure of haemopoiesis. Haemopoiesis begins with the pluri-potent haemopoietic stem cell which is capable of reconstituting blood cell production in recipient animals after irradiation or chemotherapy. This cell can either self-renew or undergo commitment to develop or differentiate into committed progenitor cells. The CFU-S (colony-forming unit – spleen) and the high proliferative potential colony-forming cell (HPP-CFC) detect cells which belong to the stem cell compartment. CFC-Mix (colony-forming cells – mixed) or CFU-GEMM (colony-forming unit – granulocyte, erythroid, megakaryo-cyte, macrophage) are cells which in semi-solid media can form multi-lineage colonies, and as such may have some overlap with those cells defined by CFU-S or HPP-CFC assays. Committed progenitor cells of several distinct haemopoietic lineages have been identified *in vitro*. These include the basophil colony-forming cell (Bas-CFC); the megakaryocyte-CFC (Meg-CFC); the granulocyte/macrophage-CFC (GM-CFC); the burst-forming unit – erythroid (BFU-R); and the eosinophil colony-forming cell (Eos-CFC). Also, pre-B lym-phocytic cells and Pre-T lymphocytic cells can be recognized in appropriate assays *in vitro*. These cells are named colony-forming cells because of their ability to form a clonal colony of mature cells of a given lineage in semi-solid culture medium *in vitro*.

1.2 The myeloid committed progenitor cell

The transit population between the stem cell compartment and mature cells has been shown to contain actively proliferating cells *in vivo* (Lajtha, 1982). When these cells are removed from the bone marrow, however, they not only cease proliferating, but also die within 24–48 hours, via a process of programmed cell death involving endonuclease activation known as apoptosis (Williams *et al.*, 1990). The onset of this process can be alleviated if the appropriate haemopoietic growth factors are added to the committed progenitor cells (Williams *et al.*, 1990). These growth factors were first shown to exist in preparations of medium conditioned by transformed cells, or by using feeder layers of these cells to provide growth factors to haemopoietic cells in the same culture. The addition of the appropriate conditioned medium to bone marrow cells in a soft gel matrix resulted in the formation of colonies (>50 cells) containing mature cells such as macrophages or granulocytes. The types of colonies formed in these soft agar assays has led to the identification of several kinds of committed progenitor cell types which can be detected in normal bone marrow (Metcalf, 1984). All these cell types are derived from the pluripotent stem cell compartment, and apparently display a fidelity to one (or two) lineages after commitment has occurred. The distinct types of committed progenitor cells are shown in Figure 1. Using the colony forming assays, it has been possible to confirm that the committed progenitor cells freshly isolated from normal bone marrow are indeed actively cycling. The stimuli for the proliferation of these cells within the bone marrow have not as yet been unequivocally identified. However, there is a good deal of evidence to suggest that extracellular matrix components, such as heparan sulphate, bind haemopoietic growth factors and can present them to the committed progenitor cells within the bone marrow (Gordon *et al.*, 1987). Some of the haemopoietic growth factors have also been shown to exist in both soluble and membrane-associated forms, so the intimate cell contact required for stromal cell-mediated haemopoietic cell development to occur may be due to presentation of growth factors via intimate cellular interactions.

2 BIOLOGICAL EFFECTS OF MYELOID GROWTH FACTORS AND THE ROLE OF G-PROTEINS IN THEIR ACTION

Many of the haemopoietic growth factors have now been identified, characterized and their amino acid sequences determined. A list of these growth factors and the haemopoietic progenitor cells that respond to each of them is shown in Table 1. The effects of haemopoietic growth factors on these target cell populations are described below, with a consideration of the possible role of G-proteins in growth factor-mediated responses in these cells.

Table 1. Growth factors and haemopoietic progenitor cells that respond to them

Growth factor or cytokine population	Responsive progenitor cell
Granulocyte macrophage colony-stimulating factor (GM-CSF)	GM-CFC (promotes mainly macrophage development) EOS-CFC BFU-E Meg-CFC
Interleukin 1 (IL-1)	Multipotent cells (synergizes with other growth factors)
Interleukin 6 (IL-6)	Multipotent cells (synergizes with other growth factors) GM-CFC (promotes development of neutrophils)
Granulocyte-CSF (G-CSF)	Multipotent cells (synergizes with other growth factors) GM-CFC (promotes development of neutrophils)
Interleukin 3 (IL-3)	Multipotent cells GM-CFC (promotes development of neutrophils and macrophages)
Macrophage-CSF (M-CSF or CSF-1)	GM-CFC (promotes development of macrophages)
Interleukin 4 (IL-4)	BFU-E, Bas-CFC (synergistically with other growth factors)

2.1 Granulocyte macrophage colony stimulating factor

Granulocyte macrophage colony stimulating factor (GM-CSF) is a pleiotropic growth factor which was originally isolated and purified on the basis of its ability to stimulate the formation of colonies containing mature macrophages and neutrophils, but a wider range of target cells is now known to respond to GM-CSF. These include eosinophilic granulocyte progenitor cells, some pluripotent haemopoietic cells, and early committed progenitor cells of the erythroid lineage (which are defined in the burst-forming unit erythroid assay), and at high concentrations GM-CSF can also promote the proliferation of megakaryocytic precursor cells (see Whetton and Dexter, 1989). These *in vitro* effects are partially mirrored in the observed effects of infusing GM-CSF into animals where an acute, dose-dependent neutrophilia is observed coupled with an increase in eosinophil and monocyte counts (Donahue *et al.*, 1986; Nienhuis *et al.*, 1987). Perhaps surprisingly there is also an increase in the number of lymphocytes present in the peripheral blood, although there is no evidence from *in vitro* experiments that GM-CSF affects lymphoid precursor cell development directly. A number of clinical trials have now been performed to investigate the ability of human GM-CSF to relieve neutropenia after chemotherapy in cancer and for some haemopoietic disorders such as myelodysplastic syndromes (Moore, 1990; Crowther *et al.*, 1990; Morstyn *et al.*, 1989). GM-CSF treatment led to an increase in circulating levels of neutrophils, eosinophils and monocytes, although some side effects such as fever and myalgia were also reported. Nonetheless there is a great deal of interest in the clinical applications of this growth factor.

The effects of GM-CSF on haemopoietic cells are not confined to the progenitor cells. Treatment of neutrophils and eosinophils with GM-CSF stimulates increased levels of functional activity. This is due to a combination of GM-CSF-stimulated maintenance of viability and activation of cytotoxic mechanisms in these cells (Begley *et al.*, 1986; Fleischmann *et al.*, 1986). GM-CSF can also stimulate the cytotoxic activity of macrophages, and the intracellular killing of parasites within macrophages, but such effects may be mediated by the release of paracrine factors such as interleukin 1, tumour necrosis factor, macrophage colony-stimulating factor and several types of interferon as well as prostaglandins and arachidonic acid (Whetton and Dexter, 1989). GM-CSF also affects the expression of cell surface adhesion molecules on myeloid cells which presumably would lead to the association of monocytes with the vascular wall and the eventual entry of these cells into the tissues (Buckle and Hogg, 1989).

Thus there are a number of functions that localized release of this cytokine may serve at sites of infection or inflammation: the attraction of haemopoietic cells (GM-CSF is a chemoattractant for myeloid cells), and adhesion and retention at specific sites; increased viability and cytotoxicity; and the stimulation of increased proliferation and chemotaxis of endothelial cells (into sites of tissue damage) (Bussolino *et al.*, 1989).

One aspect of the GM-CSF priming observed in neutrophils is the increased reponse to acute activators such as the chemotactic peptides (e.g. formyl methionine–leucine–phenylalanine, FMLP). The priming effect of GM-CSF on neutrophils is partially inhibited by pertussis toxin, implying that there may be a role for modulation of guanyl nucleotide regulatory protein activity in neutrophil priming. The GM-CSF-mediated induction of enhanced levels of platelet activating factor and chemotactic peptide induced calcium mobilization and reactive oxygen intermediate (ROI) production is inhibited by pretreating neutrophils with pertussis toxin (Mege *et al.*, 1989; McColl *et al.*, 1989). Similarly, GM-CSF-activated degranulation, release of lactoferrin and myeloperoxidase from secondary granules (Richter *et al.*, 1989), and stimulation of leukotriene synthesis (McColl *et al.*, 1989) can be partially inhibited by pertussis toxin. Membrane preparations from GM-CSF-pretreated neutrophils have higher basal and chemotactic peptide (FMLP)-stimulated GTPase activities. Pretreatment of neutrophils with pertussis toxin prior to the preparation of membranes reduces this GM-CSF enhancement of both basal and FMLP stimulated GTPase activity (Gomez-Cambronero *et al.*, 1989a). GM-CSF-mediated activation of FMLP-stimulated GTPase therefore seems to be regulated by a pertussis toxin-sensitive guanyl nucleotide regulatory protein. However, FMLP-stimulated inositol lipid hydrolysis in human neutrophils is also sensitive to pertussis toxin. The lack of effect of GM-CSF on inositol lipid hydrolysis suggests that at least two pertussis toxin-sensitive proteins exist in neutrophils, one of which can modulate GM-CSF-stimulated priming and activation of these cells. This GM-CSF-sensitive G-protein can apparently affect (directly or indirectly) a tyrosine kinase, as there is a marked inhibition of GM-CSF-stimulated tyrosine phosphorylation in pertussis toxin-treated neutrophils (Gomez-Cambronero *et al.*, 1989a,b).

Recently the low-affinity form of the human GM-CSF receptor has been cloned (Gearing *et al.*, 1989). When transfected into a murine haemopoietic cell line, low affinity binding of human GM-CSF is observed and high concentrations of hGM-CSF elicit a proliferative response. This receptor belongs to a cytokine receptor superfamily whose other members include the interleukin 3, erythropoietin, prolactin and interleukin 2 receptors (Cosman *et al.*, 1990). Although GM-CSF can stimulate tyrosine phosphorylation, as can IL-3 and IL-2, no tyrosine kinase consensus sequence has been found in any of the inferred amino acid sequences for this receptor family. A second subunit of the GM-CSF receptor can interact with this low-affinity receptor to form a high-affinity GM-CSF binding site (Hayashida *et al.*, 1990), but this second subunit also lacks a tyrosine kinase consensus sequence. Neither of these subunits bears any resemblance to the receptor family known to couple to pertussis and cholera toxin-sensitive G-proteins. The effects of pertussis toxin on GM-CSF-mediated neutrophil priming may therefore be somewhat removed from the direct effects mediated by GM-CSF receptor occupation.

2.2 Interleukin 3

Interleukin 3 (IL-3), or multi-CSF, is a glycoprotein produced by activated T lymphocytes, and possibly also keratinocytes (see Whetton and Dexter, 1990). This growth factor acts on the committed progenitor cell populations from several distinct lineages and probably has a wider range of haemopoietic target cells than any other cytokines acting on myeloid cells (see Table 1). IL-3 can stimulate the proliferation and development of multipotent cells identified using the CFU-S, the high proliferative potential colony-forming cell (HPP-CFC) and the mixed colony-forming cell (CFC-Mix) assays (Heyworth *et al.*, 1988). The committed progenitor cells from several myeloid lineages are also capable of responding to IL-3: these include megakaryocytic progenitor cells, mast cell progenitor cells, granulocyte/macrophage colony-forming cells, erythroid progenitor cells at both early (BFU-E) and late (CFU-E) stages in their development, and eosinophil precursor cells (Metcalf, 1984; Whetton and Dexter, 1989). There is also evidence that cells capable of forming T and B lymphocytes respectively can respond to IL-3 (Clark-Lewis and Schrader, 1988). These data suggest that primitive haemopoietic cells which are not committed to either lymphoid or myeloid development are responsive to IL-3. The profound effects of this growth factor on early progenitor cells have also been demonstrated in experiments where IL-3 has been infused into mice; this leads to massive increases in the committed progenitor cells within the haemopoietic organs (Kindler *et al.*, 1986; Lord *et al.*, 1986), although the peripheral blood levels of myeloid cells remain relatively unchanged. The infusion of low levels of GM-CSF after seven days treatment with IL-3 can stimulate a significant increase in the numbers of eosinophils, basophils, lymphocytes and neutrophils present compared with animals treated with GM-CSF alone (Donahue *et al.*, 1988). This suggests that the *in vivo* effects of IL-3 are predominantly on primitive cells, and the further infusion of GM-CSF can act on the progenitor cells formed in response to IL-3, leading to increased levels of peripheral blood cells from several lineages.

At present few data are available concerning IL-3-stimulated signal transduction mechanisms. Specific cell surface receptors have been identified which (on bone marrow cells) have a K_D of about 200 pM (Nicola and Metcalf, 1986). The number of receptors expressed per cell is only about 120, and doses of IL-3 which elicit a complete biological response will only achieve $<20\%$ receptor occupation. Clearly there must be a significant degree of amplification of cellular signalling events after the IL-3 receptor is occupied. Recently a low-affinity form of the IL-3 receptor has been cloned which is a member of the cytokine receptor superfamily (Schreurs *et al.*, 1990; Itoh *et al.*, 1990). Whether this low-affinity form associates with other proteins (as does the GM-CSF low-affinity receptor, see above) remains unclear. There is some evidence that IL-3 stimulates proliferation via a mechanism involving a pertussis toxin-sensitive G-protein. The incubation of an IL-3-dependent cell line, FDCP-1, with pertussis toxin can partially inhibit IL-3-stimulated proliferation (He *et al.*, 1988; Kelvin *et al.*, 1989). This may be associated with the ability of pertussis toxin to increase the cell cycle length of IL-3 dependent cells (Koyasu *et al.*, 1989). However, other biological and biochemical responses to IL-3 are also modified in pertussis toxin-treated cells; IL-3-stimulated hexose transport and Na^+/H^+ antiport activation are both inhibited by pretreatment of the cells with pertussis toxin (Kelvin *et al.*, 1989; A.A. Akinsanya and A.D. Whetton, unpublished observations). At present very little is known of the mitogenic signals elicited by the occupied IL-3 receptor (see Vallance and Whetton, 1991); it does appear, however, that these signalling events may involve a pertussis toxin-sensitive G-protein. Whether this G-protein couples directly to the IL-3 receptor remains to be seen.

2.3 Granulocyte colony-stimulating factor

Granulocyte colony stimulating factor (G-CSF) is a pleiotropic cytokine which was originally identified and purified by its ability to induce the differentiation of leukaemic cell lines *in vitro*, and to promote the development of bone marrow progenitor cells to form colonies which consist of neutrophils. Like GM-CSF there is also an ability to promote survival and limited proliferation in early erythroid and eosinophilic progenitor cells. This may in part be related to the observation that G-CSF can shorten the Go period of the multipotent cells found in normal bone marrow. When G-CSF and IL-3 are added to soft gel cultures containing primitive multipotent cells, they act synergistically to promote the development of colonies containing mature cells from several different lineages (Ikebuchi *et al.*, 1988; Heyworth *et al.*, 1988). These data imply that G-CSF can act at several different levels of haemopoietic cell development. But, like GM-CSF, G-CSF can also affect the *functional* activity of neutrophilic granulocytes. G-CSF can increase the lifetime of circulating neutrophils, and enhance their antibody-dependent cellular cytotoxicity (Begley *et al.*, 1986; Lopez *et al.*, 1983). As there is a dramatic elevation in serum levels of G-CSF in bacterially infected mice, G-CSF (and GM-CSF) probably have a role in host defence mechanisms via the ability to stimulate myeloid cell production and prime mature myeloid cell functional activity. G-CSF also promotes the proliferation of endothelial cells, suggesting that localized release of this cytokine at sites of infection or lesions in the endothelium may promote regeneration of damaged tissue (Bussolino *et al.*, 1989).

When infused into patients or animals, G-CSF can rapidly elevate leukocytenumbers and activate circulating neutrophils (Morstyn *et al.*, 1989; Whetton and Dexter, 1989). The lack of major side effects of this cytokine, coupled with its ability to profoundly decrease the rate of infections in neutropenic patients (e.g. after chemotherapy, or during haemopoietic recovery after irradiation), suggests that this cytokine will have a wide variety of clinical applications (Morstyn *et al.*, 1989; Moore, 1990). Nevertheless, little is known about the molecular mechanisms whereby G-CSF achieves these profound effects on haemopoietic cells. A G-CSF receptor has recently been cloned and its amino acid sequence derived (Fukunaga *et al.*, 1990), but there are few data available on the events triggered immediately after receptor occupation. G-CSF-treated membranes from the leukaemic NFS-60 cell line do exhibit an increased affinity for the non-hydrolysable GTP analogue GTPγS, and elevated levels of adenylyl cyclase activity (Matsuda *et al.*, 1989). However, the precise role of adenylyl cyclase in the mode of action of G-CSF awaits further investigation.

2.4 Macrophage colony-stimulating factor

Unlike the other myeloid growth factors, macrophage colony-stimulating factor (M-CSF or CSF-1) is a homodimer, disulphide bridges joining the two monomers to form the biologically active growth factor. Differential mRNA processing can give distinct forms of M-CSF, one of which is membrane bound (Wong *et al.*, 1988). M-CSF stimulates the survival and proliferation of monocytes and macrophages, and their progenitor cells in the haemopoietic tissues. There is also evidence of a role for M-CSF in placental development (Sherr, 1990), although the physiological importance of this has yet to be established. In soft agar cultures of normal bone marrow cells, M-CSF will stimulate the formation of macrophages from the bipotential granulocyte macrophage colony-forming cells. This process requires the presence of serum, suggesting a requirement for a secondary, as yet unidentified, growth factor to stimulate macrophage formation in concert with M-CSF. The response of multipotent cells to M-CSF is also limited unless a second growth factor is present, but in this case the agent required has been identified as IL-1 (or IL-6) (Mochizuki *et al.*, 1987; Heyworth *et al.*, 1988; Ikebuchi *et al.*, 1988). The addition of either M-CSF or IL-1 alone to multipotent cells cannot stimulate development, but when they are added together these two agents synergistically promote the development of large numbers of mature macrophages. The molecular basis for this synergy is as yet unclear.

Like G-CSF and GM-CSF, M-CSF can act on mature cells to enhance levels of mature cell function. M-CSF supports the survival and proliferation of tissue based macrophages and peripheral blood monocytes. Furthermore, it can prime these cells to display increased levels of prostaglandin, GM-CSF, interferon and tumour necrosis factor synthesis. Similarly, enhanced levels of superoxide production, tumour cell killing and phagocytosis are observed in monocytes and macrophages that have been pretreated with M-CSF.

M-CSF elicits its effects on the responsive cells described above by binding to a single class of high-affinity receptors, encoded by the proto-oncogene c-*fms*. The M-CSF receptor is a 160-kDa protein which is structurally related to the PDGF receptor and consists of an extracellular ligand-binding domain, a membrane-spanning segment and an intracellular enzymic domain that specifies tyrosine specific protein kinase activity (Sherr, 1990). M-CSF can stimulate receptor oligomerization, leading to the activation of the tyrosine kinase and the tyrosine phosphorylation of a number of cytosolic and membrane-bound proteins (Sengupta *et al.*, 1988; Downing *et al.*, 1988). The relevant physiological substrates for the M-CSF receptor which transduce the mitogenic effects of this growth factor still remain to be identified, but recently c-*raf* has been identified as one possible substrate, as it is for the related PDGF receptor (Baccarini *et al.*, 1990). However, another PDGF receptor substrate, phosphatidylinositol 4,5-bisphosphate-phospholipase Cγ, is not phosphorylated in response to M-CSF, suggesting that differences in the mode of action of these two growth factors exist.

One possible difference lies in the role of pertussis toxin-sensitive G-proteins in the mode of action of these two growth factors. M-CSF can stimulate an increase in the affinity of human monocyte membranes for the non-hydrolysable GTP analogue GTPγS, and furthermore induces a stimulation of GTPase activity in monocyte membranes which is sensitive to pertussis toxin (Imamura and Kufe, 1988). This M-CSF-stimulated activation of a pertussis toxin-sensitive G-protein may be associated with some of the signal transduction events mediated by this growth factor, as M-CSF-stimulated Na$^+$/H$^+$ antiport activation and Na$^+$ influx are inhibited by pre-incubation with pertussis toxin. M-CSF can also stimulate phosphatidylcholine hydrolysis, thereby activating protein kinase C, events which are inhibited by pertussis toxin (Imamura *et al.*, 1990). These effects of pertussis toxin on M-CSF-stimulated signalling events are not limited to monocytes; when c-*fms* was introduced into hamster fibroblasts they could be stimulated to proliferate in the presence of M-CSF, and M-CSF- (but not PDGF-) stimulated DNA synthesis and Na$^+$/H$^+$ antiport activation were partially inhibited by pertussis toxin (Hartmann *et al.*, 1990). The possible mechanisms whereby the M-CSF receptor kinase might influence the activity of a pertussis toxin-sensitive G-protein remain to be established.

2.5 ADP-ribosylation inhibitors and development

Some of the evidence described above suggests that there is a role for G-proteins in the control of proliferation and development of myeloid haemopoietic cells mediated by growth factors. A slightly different approach to this question has been the use of inhibitors of cholera and pertussis toxin-stimulated ADP-ribosylation on haemopoietic cell proliferation and development in long-term bone marrow cultures. These cultures consist of bone marrow stromal cell layers in close association with haemopoietic cells. The stromal cells promote the proliferation and differentiation of multipotent cells to form firstly committed progenitor cells, and eventually mature haemopoietic cells. Under specific conditions these cultures produce mainly neutrophils and macrophages from multipotent cells. However, on addition of inhibitors of monoADP-ribosyltransferases there is a complete cessation of normal myeloid development: the number of multipotent cells remains unaltered but the production of GM-CFC and neutrophils is markedly depleted (Dexter *et al.*, 1985). The monoADP-ribosyltransferase inhibitors apparently block the differentiation of multipotent cells to form lineage-restricted committed progenitor cells in long-term bone marrow cultures, and this effect is also observed in colony-forming assays when IL-3 is employed as the growth stimulus (Heyworth and Dexter, 1988). PolyADP-ribosylation inhibitors have no such effect. Whilst these studies imply a role for monoADP-ribosylation and thus possibly G-proteins in development, a greater understanding of haemopoietic cell differentiation and development will be required to understand the significance of these observations.

3 MYELOID LEUKAEMIAS

The apparent role of G-proteins in the control of proliferation and development of haemopoietic cells suggests that the disregulated blood cell production seen in leukaemia may be associated with the oncogenic G-protein *ras*. There is now considerable evidence that this is indeed the case.

Leukaemia is a neoplastic disorder of the blood which is clonal in nature and which originates in the haemopoietic stem cell compartment. There are several forms of leukaemia which affect different blood cell lineages. In addition to the true leukaemias, there is a group of blood disorders collectively termed the myelodysplastic syndromes (MDS), which represent pre-leukaemic conditions involving abnormal blood cell production that may eventually evolve into leukaemia. Of the leukaemias, there are two broad types; the acute leukaemias in which onset is sudden and disease progression rapid, and the chronic forms which show a more gradual and prolonged pathology.

Acute myeloid leukaemia (AML) and acute lymphoblastic leukaemia (ALL) have several clinical features in common at presentation, all of which are a consequence of bone marrow failure (i.e. to produce normal mature haemopoietic cells), and the over-production and accumulation of immature leukaemic cells in the blood. AML collectively refers to seven disease subtypes distinguished by the degree to which the leukaemic cells are differentiated. The different subtypes (M_1–M_7) are morphologically highly varied; they may involve production of completely undifferentiated blast cells as in the myeloblastic M_1 subtype, or alternatively give rise to relatively well differentiated cells such as the erythroblasts seen in M_6 erythroleukaemia.

In most cases the causes behind leukaemia remain unknown, but some role for mutational activation of *ras* genes as a causative agent in leukaemogenesis has been inferred. *Ras* genes have been shown to transform haemopoietic cells *in vitro*, particularly those of myeloid lineages, and render them growth factor independent (Andrejauskas and Moroni, 1989; Boswell *et al.*, 1990; Nair *et al.*, 1989). However, blast cells isolated from patients with acute leukaemia still require colony-stimulating factors for their continued survival and proliferation *in vitro*, which implies that such studies are of little relevance to leukaemogenesis *in vivo*. Rather than autonomous growth of immature blood cells, acute myeloid leukaemia seems to be a consequence of a block in the differentiation of blood cell progenitors, which remain growth factor dependent. Thus the observation of a differentiation block induced by *ras* in a myeloid cell line which remained G-CSF dependent is of particular relevance, since it suggests *ras* activation may be associated with the inhibition of development of leukaemic myeloid progenitor cells (Mavilio *et al.*, 1989). Although overexpression of *ras* genes can induce transformation, this does not appear to be a common clinical event in leukaemias. However, mutations within *ras* genes clustered around codons 12/13 (and also 11) and in codon 61 can also activate their oncogenic potential, and it is these mutations which are frequently seen in various leukaemic conditions (Bartram, 1988). Mutations are most frequently seen in N-*ras*; those in K-*ras* are rare, and those in H-*ras* rarer still.

4   *RAS* GENES IN THE ACUTE LEUKAEMIAS AND MYELODYSPLASIAS

Ras mutations occur most frequently in patients presenting with AML, although a substantial number of patients with ALL and MDS also exhibit *ras* mutations. The exact percentage of cases that do harbour a mutated *ras* allele varies substantially between studies performed on different groups of patients; the different subtypes of AML and the use of patients at different stages of the disease further complicate these studies, but data pooled from a number of investigations suggest an incidence of 30–36% (Bartram, 1988; Butturini and Gale, 1990) for the occurrence of *ras* mutations in AML. Of these *ras* mutations, 90% were found to be in N-*ras*. Similarly, of the *ras* mutations identified in 27% of AML patients by Farr *et al.* (1988), all were found to be in N-*ras*, predominantly at codon 12. N-*ras* mutations at codons 12 and 13 occur with equal frequency, but, interestingly, very few patients appear to have a mutation in codon 61 (Ahuja *et al.*, 1990). Mutations in H-*ras* in myelodysplasias have been detected (Gow *et al.*, 1988; Padua *et al.*, 1988), but they are very rare in the acute leukaemias (Browett *et al.*, 1989).

Although long-term survival in AML is rare, intensive therapy can induce clinical remission, defined as when blood cell counts normalize, blast cells represent less than 5% of the total marrow cell population, and extramedullary evidence of the disease is absent, but this does not necessarily represent total eradication of leukaemic cells from the body. In cases where mutated *ras* genes have been found at onset and remission occurs, peripheral blood cells show no evidence of mutations in these *ras* genes. It appears that the normal phenotype re-emerges and that *ras* mutations are specifically associated with the leukaemic cell population. The loss of mutations in N-*ras* from the blood cells of patients has been observed in cases of remission from AML (Senn *et al.*, 1988b; Saglio *et al.*, 1989), as well as from acute non-lymphocytic leukaemia (ANLL) (Senn *et al.*, 1988a). In the latter case, follow-up studies of three patients demonstrated the absence of the N-*ras* mutation previously identified following successful induction of remission by chemotherapy. In contrast, this mutation persisted in the third patient, who failed to achieve remission. Of the N-*ras* mutations found in 25% of untreated AML patients sutdied at onset by Saglio *et al.* (1989), none were present in remission samples. However, of four patients subsequently entering a complete remission, neither of the two that previously had a mutated *ras* gene at disease onset were found to have kept this aberration. In contrast, two patients who underwent relapse maintained the same *ras* mutation diagnosed at onset. It is possible that relapse occurs as a result of residual leukaemic cells which may already harbour a mutated *ras* gene re-entering the cell cycle, but the importance of *ras* mutations in relapse is unclear. The re-emergence of blood cells with a *ras* mutation has been observed in relapse (Saglio *et al.*, 1989), but in other studies the relapse cell population does not show *ras* mutations in cases where they were present at onset (Senn *et al.*, 1988b; Farr *et al.*, 1988). Thus it would seem that relapse may involve re-emergence of the leukaemic clone in some cases, but that it may also represent the induction of a new leukaemia *per se*. Several surveys involving large numbers of patients with acute leukaemias have also underlined the importance of *ras* mutations in ALL and myelodysplasias, particularly in the progression of the latter into AML (Browett and Norton, 1989; Ahuja *et al.*, 1990).

The myelodysplasias are a group of chronic myeloproliferative disorders broadly divided into two types; refractory anaemias that show excess blast production, and chronic myelomonocytic leukaemias. Both types are chronic in nature, and both are characterized by a hypercellular bone marrow containing excessive blasts; mature neutrophils in the peripheral blood may be abnormal. MDS eventually undergoes transformation from a relatively benign chronic phase lasting between months and years, into acute myeloid leukaemia, and there is considerable evidence that activation of *ras* genes may have a role to play in the change in disease pathology. Several studies reveal a very high incidence of *ras* mutations in MDS patients (Yunis *et al.*, 1989; Ahuja *et al.*, 1990; Padua *et al.*, 1988), even more profound in patients undergoing progression into AML (Yunis *et al.*, 1989); 41% of patients were found to have mutations in *ras* genes, mainly in N-*ras*. Of these, 73% were in progression of the disease into AML. Furthermore, 82% of patients with *ras* mutations were found to have disrupted monocytic cell compartments whereas lymphoid cell production was relatively unaffected. This suggests that *ras* mutations may affect myeloid cell development rather more profoundly than lymphopoiesis. This is borne out by the observed incidence of *ras* mutations in acute leukaemias being highest in myeloid disorders such as AML and MDS as compared to lymphoid malignancies, e.g. ALL. Importantly, of these MDS patients, those with chronic myelomonocytic leukaemia (CMML) showed a particular predisposition to *ras* mutation, as has previously been observed (Cogswell *et al.*, 1989). Evidence for any role of overexpression as opposed to mutation of *ras* in leukaemias is scant; mutational activation of *ras* genes is by far the most common aberration, although a six-fold increase in K-*ras* expression has been observed in myelodysplasia (Srivastava *et al.*, 1988).

5 LEUKAEMIA AND THE PHILADELPHIA CHROMOSOME

Chronic myeloid leukaemia (CML) is a haemopoietic disorder first documented in the mid-nineteenth century; the disease is generally fatal despite a variety of treatments, and in the vast majority of cases the causative agent(s) remains unknown. An initial chronic phase, lasting up to several years, sees the elevated production of committed myeloid precursors, generally granulocytic in nature, but in some cases thrombocytes (Champlin and Golde, 1985). Typically, granulocytes accumulate in the peripheral blood, together with their immature precursor cells, and more than 95% of CML patients display the cytogenetic abnormality termed the Philadelphia (Ph[1]) chromosome. This was originally described as a small chromosome 22 (Nowell and Hungerford, 1960), since when CML has become the first and most well documented human neoplasm to be consistently associated with a chromosomal abnormality. Use of enzyme marker analysis (Fialkow *et al.*, 1977) and detailed karyotype analysis has shown the disease to be clonal in nature, arising from the transformation of a pluripotent haemopoietic stem cell.

The foreshortened long arm chracteristic of the Philadelphia chromosome results from the reciprocal translocation of genetic material from chromosome 9(q34) to the long arm of chromosome 22. The t(9: 22)(q34:q11) translocation moves a large portion of the c-*abl* proto-oncogene from chromosome 9 and relocates it in a 5.8-kb region on chromosome 22 known as the breakpoint cluster region (bcr) (Grosveld *et al.*, 1986; Groffen *et al.*, 1984). This bcr is located within a large gene called BCR which normally encodes a phosphoprotein of 160–190 kDa, whose function remains unknown. Juxtaposition of the c-*abl* and BCR genes results in the formation of a chimeric bcr–abl gene encoding a 210-kDa polypeptide, p210$^{bcr-abl}$ (Shtivelman *et al.*, 1985). An important feature of p210$^{bcr-abl}$ is the deletion of N-terminal residues and their replacement by residues from the *BCR* gene (Ramakrishnan and Rosenberg, 1989; Risser and Holland, 1989). The Philadelphia chromosome has also been observed in patients with ALL; around 20% of adult ALL patients are Ph[1]-positive (Priest *et al.*, 1980), and some cases of Ph[1]-positive ALL are associated with the expression of a 190-kDa bcr–abl fusion protein generated by a breakpoint within exon 1 of BCR (Hermans *et al.*, 1987). It is also possible that some Ph[1]-positive cases of ALL which show rearrangement within the bcr represent cases of

CML in lymphoid blast crisis following a chronic phase which has escaped diagnosis.

Leukaemic Ph^1-positive cells clearly have a growth advantage over their normal counterparts, since the proportion of Ph^1-positive cells increases throughout the chronic phase until they constitute virtually 100% of the blood cell population, although normal Ph^1-negative cells are still present (Singer *et al.*, 1980). Importantly, the precursor cells produced during the chronic phase are still capable of terminal differentiation, and are growth factor dependent (Chervenick *et al.*, 1971; Metcalf *et al.*, 1974). Chronic-phase granulocytosis is now thought to be due to accelerated expansion and maturation of committed progenitors, rather than uncontrolled proliferation of Ph^1-positive stem cells into a particular lineage (Strife *et al.*, 1988).

For reasons which remain unclear, CML in chronic phase undergoes transformation into an aggressive acute phase, which in certain aspects resembles the acute leukaemias (see above). This so-called blast crisis results from the rapid clonal outgrowth of primitive blast cells and promyelocytes from an immature Ph^1-positive subclone, which can be of myeloid (in around 70% of cases) or lymphoid lineage. Myeloid blast crisis generally sees outgrowth of myeloblast-like cells, although more rarely erythroid or megakaryocytic blast cells predominate (Bain *et al.*, 1977; Rosenthal *et al.*, 1977). The appearance of blast cells of different lineages is an important part of the concept that CML originates from a single clone of haemopoietic stem cells. Primitive blast cells produced in acute phase are in some way under developmental arrest, since they will no longer undergo terminal differentiation (Golde *et al.*, 1974). The appearance of secondary chromosomal aberrations at the onset of blast crisis is common (Champlin and Golde, 1985) and may be associated with the profound change of course in the disease at this stage.

5.1 The bcr–abl gene product

Both $p210^{bcr-abl}$ and $p190^{bcr-abl}$ are tyrosine kinases, and both will transform haemopoietic cells *in vitro*. Significantly, the $p185$-$190^{bcr-abl}$ gene associated with ALL was found to have more transforming activity than its CML-specific $p210$ counterpart, and this has been correlated with its higher levels of tyrosine kinase activity. However, the use of bcr–abl transfected bone marrow to reconstitute mice, and indeed mice transgenic for a bcr–abl gene, have recently demonstrated the induction of haemopathies by the bcr–abl gene product. Prominent amongst a range of disorders induced by p210$^{bcr-abl}$ was a myeloproliferative syndrome which closely resembled chronic-phase CML in humans (Daley *et al.*, 1990), and there was even some evidence of a lymphoid blast crisis in one of the reconstituted mice. A myeloproliferative syndrome was also observed by Kelliher *et al.* (1990), where several features of CML were prevalent, namely clonal outgrowth of bcr–abl-positive granulocytic cells, and granulocytic infiltration of the spleen. Elefanty *et al.* (1990) induced a variety of haemopoietic neoplasms with a bcr–abl gene which involved both myeloid and lymphoid cells, although only a mild CML-like condition was seen in some recipients. Mice transgenic for a bcr–v-abl oncogene were found to develop T cell and pre-B cell lymphomas (Hariharan *et al.*, 1989), but transgenic mice carrying a p190$^{bcr-abl}$ gene were found to develop acute myeloid or lymphoid leukaemia (Heisterkamp *et al.*, 1990). Although the importance of the bcr–abl genes in the pathogenesis of CML in particular has become evident, its precise role in the disease remains unclear. Furthermore, the events underlying the transition of CML in chronic phase to blast crisis are also largely unresolved. However, in around 80% of CML patients in acute phase, the occurrence of further gene rearrangements and genetic abnormalities can be detected, and it is now thought that these may have an important role to play in disease transformation.

5.2 Secondary genetic changes and blast crisis in CML

The high incidence of secondary genetic changes observed upon the onset of blast crisis suggests that these may have some role in causing disease transformation. Such changes might lead to the activation of other oncogenes which are then responsible for the changes in pathology characteristic of acute-phase CML. Of particular interest in this respect are the *ras* genes, since their activation by point mutation has been demonstrated in a variety of human leukaemias, particularly in aggressive acute leukaemias, e.g. AML and ALL. The idea that the activation of *ras* genes by secondary mutations may precipitate the transformation of chronic-phase CML into blast crisis has been investigated in some depth.

Although activation of *ras* genes provides an attractive model for the transformation of chronic phase into blast crisis, *ras* mutations appear to be rare events in the disease. Although *ras* mutations are present in some patients with CML during the acute phase, such mutations occur too infrequently to have a significant role in acute-phase onset in general. *Ras* mutations have been demonstrated in patients in both stages of CML; activation of H-*ras* has been observed in blast crisis (Gow *et al.*, 1988), and in another study only one of six patients in chronic phase harboured a *ras* mutation, as opposed to three of six in blast crisis (Liu *et al.*, 1988). In contrast, none of 12 patients with bcr–abl-positive CML in blast crisis showed activation of either N- or H-*ras* (Naoe *et al.*, 1989), and no *ras* mutations were found in any of 26 cases of CML examined by Janssen *et al.* (1987), more than half of which were in acute phase. A large number of studies failed to demonstrate a correlation between *ras* activation and onset of acute phase. No *ras* mutations were seen in 44 cases of CML, either in chronic or acute phase, except in two patients with a K-*ras* mutation upon transformation into myeloid blast crisis (LeMaistre *et al.*, 1989). However, only one of these was shown to lack this mutation in chronic phase. Similarly, no *ras* activation was observed in any of 44 patients with CML examined by Collins *et al.* (1989), nearly half of whom were in blast crisis. Several other studies have further reinforced the idea that, overall, *ras* mutations in CML are rare, and if present at all prevail in the latter stages of the disease (Saglio *et al.*, 1989; Ahuja *et al.*, 1990; Needleman *et al.*, 1989). Interestingly, one examination of CML patients has shown a high incidence of *ras* mutation in atypical CML. None of the patients examined with classical, i.e. bcr–abl rearrangement positive CML in chronic phase, and only one of 18 patients in blast crisis, showed a mutated *ras* gene. However, 54% of the patients with bcr–abl-negative CML did indeed harbour a *ras* mutation (Cogswell *et al.*, 1989). This particular form of CML is now regarded as a myelodysplasia termed chronic myelomonocytic leu-kaemia (CMML), and provides further evidence of *ras* activation in the myelodysplastic syndromes. Overall, mutational activations of *ras* genes in CML are rare, late-stage events, which may well be associated with, rather than a cause of, chronic phase transformation. Thus, although there is now some evidence that the progression of CML into blast crisis is loosely associated with *ras* mutations, it is clear that they have an important role in the pathology of AML and the myelodysplasias. The exact role of these G-proteins in leukaemogenesis remains to be elucidated. The exact role of G-proteins in non-trans-formed haemopoietic cells is also unclear at present, but the few studies performed thus far do indicate that this group of proteins may have a limited role in controlling development of both stem cells and committed progenitors.

REFERENCES

Ahuja, H.G., Foti, A., Bar-Eli, M. and Cline, M.J. (1990). *Blood* **75**, 1684–1690.
Andrejauskas, E. and Moroni, C. (1989). *EMBO J.* **8**, 2575–2581.
Bain, B., Catovsky, D., O'Brien, M., Spiers, A.S.D. and Richards, H.G.H. (1977). *J. Clin. Pathol.* **30**, 235–242.
Baccarini, M., Sabatini, D.M., App, H., Rapp, U.R. and Stanley, E.R. (1990). *EMBO J.* **9**, 3649–3657.
Bartram, C.R. (1988). *Blood Cells* **14**, 533–538.
Begley, C.G., Lopez, A.F., Nicola, N.A., Warren, D.J., Vadas, M.A., Sanderson, C.J. and Metcalf, D. (1986). *Blood* **68**, 162–166.
Boswell, H.S., Nahreini, T.S., Burgess, G.S., Srivastava, A., Gabig, T.G., Inhorn, L., Srour, E.F. and Harrington, M.A. (1990). *Exp. Hematol.* **18**, 452–460.
Browett, P.J. and Norton, J.D. (1989). *Oncogene* **4**, 1029–1036.
Browett, P.J., Yaxley, J.C. and Norton, J.D. (1989). *Leukemia* **3**, 86–88.
Buckle, A.M. and Hogg, N. (1989). *J. Immunol.* **143**, 2295–2301.
Bussolino, F., Wang, J. H., DeFilippi, P., Turrini, F., Sanavio, F., Egdell, C-J.S., Aglietta, M., Arese, P. and Mantovani, :A. (1989). *Nature* **337**, 471–473.
Butturini, A. and Gale, R. P. (1990). *Leukemia* **4**, 138–160.
Byron, J.W. (1975). *Exp. Hematol.* **3**, 44–49.
Byron, J.W. (1977). *Agents Actions* **7**, 209–215.
Champlin, R.E. and Golde, D.W. (1985). *Blood* **65**, 1039–1047.
Chervenick, P.A., Ellis, C.D., Pan, S.F. and Lawson, A.L. (1971). *Science* **174**, 1134–1136.
Clark-Lewis, I. and Schrader, J.W. (1988). *Lymphokines* **15**, 1–37.
Cogswell, P.C., Morgan, R., Dunn, M., Neubauer, A., Nelson, P., Poland-Johnston, N.K., Sandberg, A.A. and Liu, E. (1989). *Blood* **74**, 2625–2633.
Collins, S.J., Howard, M., Andrews, D.F., Agura, E. and Radich, J. (1989). *Blood* **73**, 1028–1032.
Cosman, D., Lyman, S.D., Idzerda, R.L., Beckman, M.P., Park, L.S., Goodwin, R.G. and March, C.J. (1990). *Trends Biochem. Sci.* **15**, 265–269.
Crowther, D., Scarffe, J.H., Howell, A., Thatcher, N., Bronchud, M., Steward, W.P., Testa, N. and Dexter, M. (1990). In: *Molecular Control of Haemopoiesis* (CIBA Foundation Symposium 148), pp. 201–214. J Wiley, London.
Daley, G.W., VanEtten, R.A. and Baltimore, D. (1990). *Science* **247**, 824–829.
Dexter, T.M., Whetton, A.D. and Heyworth, C. M. (1985). *Blood* **65**, 1544–1588.
Donahue, R.E., Wang, E.A., Stone, D.K., Kamen, R., Wong, G.G., Sehgal, P.K., Nathan, D.G. and Clark, S.C. (1986). *Nature* **321**, 872–874.
Donahue, R.E., Seehra, J., Metzger, M., Lefebure, D., Rock, B., Carbone, S., Nathan, D.G., Garnick, M., Sehgal, P.K., Laston, D., Lavallie, E., McCoy, J., Schendel, P.F., Norton, C., Turner, K., Yang, Y-C. and Clark, S.C. (1988). *Science* **241**, 1820–1823.
Downing, J.R., Rettenmier, C.W. and Sherr, C.J. (1988). *Mol. Cell. Biol.* **8**, 1795–1799.
Elefanty, A.G., Hariharan, I.K. and Cory, S. (1990). *EMBO J.* **9**, 1069–1078.

Farr, C.J., Saiki, R.K., Erlich, H.A., McCormick, F. and Marshall, C.J. (1988). *Proc. Natl. Acad. Sci. USA* **85**, 1629–1633.

Fialkow, P.J., Jacobson, R.J. and Papayannopoulou, T. (1977). *Am. J. Med.* **63**, 125–130.

Fleischmann, J., Golde, D.W., Weisbart, R.G. and Gasson, J.C. (1986). *Blood* **68**, 708–711.

Fukunaga, R., Ishizaka-Ikeda, E., Seto, Y. and Nagata, S. (1990). *Cell* **61**, 341–350.

Gearing, D.P., King, J.A., Gough, N.M. and Nicola, N.A. (1989). *EMBO J.* **8**, 3667–3676.

Golde, D.W., Byers, L.A. and Cline, M.J. (1974). *Cancer Res.* **34**, 419–423.

Gomez-Cambronero, J., Yamazaki, M., Metwally, F., Molski, T.F., Bonak, V.A., Huang, C.K., Becker, E.L. and Sha'afi, R.I. (1989a). *Proc. Natl. Acad. Sci. USA* **86**, 3569–3573.

Gomez-Cambronero, J., Huang, C.K., Bonak, U.A., Wang, E., Casnellie, J.E., Shiraishi, T. and Sha'afi, R.I. (1989b). *Biochem. Biophys. Res. Commun.* **162**, 1478–1485.

Gordon, M.Y., Riley, G.P., Watt, S.M. and Greaves, M.F. (1987). *Nature* **326**, 403–405.

Gow, T., Hughes, D., Farr, C., Hamblin, T., Brown, R. and Padua, R.A. (1988). *Leukemia Res.* **12**, 805–810.

Groffen, J., Stephenson, J.R., Heisterkamp, N., DeKlein, A., Bartram, C.R. and Grosveld, G. (1984). *Cell* **36**, 93–99.

Grosveld, G., Verwoerd, T., Van Agthoven, T., DeKlein, A., Ramachandran, K.L., Heisterkamp, N., Stam, K. and Groffen, J. (1986). *Mol. Cell. Biol.* **6**, 607–616.

Hariharan, I.K., Harris, A. W., Crawford, M., Abud, H., Webb, E., Cory, S. and Adams, J.M. (1989). *Mol. Cell. Biol.* **9**, 2798–2805.

Hartman, T., Seuwen, K., Roussel, M.F., Sherr, C.J. and Pouyssegur, J. (1990). *Growth Factors* **2**, 289–300.

Hayashida, K., Kitamura, T., Gorman, D.M., Arai, K-I., Yokota, T. and Miyajima, A. (1990). *Proc. Natl. Acad. Sci. USA* **87**, 9655–9659.

He, Y.X., Hewlett, E., Temeles, D. and Quesenberry, P. (1988). *Blood* **71**, 1187–1195.

Heisterkamp, N., Jenster, G., ten Hoeve, J., Zovich, D., Pattengale, P.K. and Groffen, J. (1990). *Nature* **344**, 251–253.

Hermans, A., Heisterkamp, N., von Lindern, M., van Baal, S., Meijer, D., van der Plas, D., Wiedemann, L. M., Groffen, J., Bootsma, D. and Grosveld, G. (1987). *Cell* **51**, 33–40.

Heyworth, C.M. and Dexter, T.M. (1988). *Leukemia* **2**, 6–11.

Heyworth, C.M., Ponting, I.L.O. and Dexter, T.M. (1988). *J. Cell. Sci.* **91**, 239–247.

Ikebuchi, K., Ihle, J., Hirai, Y., Wong, G.G., Clark, S.C. and Ogawa, M. (1988). *Blood* **72**, 2007–2014.

Imamura, K. and Kufe, D. (1988). *J. Biol. Chem.* **263**, 14093–14098.

Imamura, K., Dianoux, A., Nakamura, T. and Kufe, D. (1990). *EMBO J.* **9**, 2423–2429.

Itoh, N., Yonehara, S., Schreurs, J., Gorman, D.M., Maruyama, K., Ai, I., Yahara, I., Arai, K-I. and Miyajima, A. (1990). *Science* **247**, 324–327.

Janssen, J.W.G., Steenvoorden, A.C.M., Lyons, J., Anger, B., Bohlke, J.V., Bos, J.L., Seliger, H. and Bartram, C.R. (1987). *Proc. Natl. Acad. Sci. USA* **84**, 9228–9232.

Kelliher, M.A., McLaughlin, J., Witte, O.N. and Rosenberg, N. (1990). *Proc. Natl. Acad. Sci. USA* **87**, 6649–6653.

Kelvin, D., Shreeve, M., McAuley, C., McLeod, D.L., Simard G., Connolly, J.A. (1989). *J. Cell. Physiol.* **138**, 273–280.

Kindler, V., Thorens, B., Kossodo, S.D., Allet, B., Eliason, J.F., Thatcher, D., Farber, N. and Vassalli, P. (1986). *Proc. Natl. Acad. Sci. USA* **83**, 1001–1005.

Koyasu, S., Miyajima, A., Arai, K., Okajima, F., Ui, M. and Yahara, I. (1989). *Cell. Struct. Function* **14**, 459–471.

Lajtha, L.G. (1982). In: *Blood and its Disorders* (Hardisty, R.M. and Weatherall, D.J. eds), pp. 57–74. Blackwell, London.

Lee, J., Talpaz, M., Walters, R., Gutterman, J.U. and Blick, M. (1988). *Hematol. Pathol.* **2**, 229–237.

Liu, E., Hjelle, B. and Bishop, J.M. (1988). *Proc. Natl. Acad. Sci. USA* **85**, 1952–1956.

Lopez, A.F., Nicola, N.A., Burgess, A.W., Metcalf, D., Battye, F., Sewell, W.A. and Vadas, M. (1983). *J. Immunol.* **131**, 2983–2988.

Lord, B.I., Molineux, G., Testa, N.G., Kelly, M., Spooncer, E. and Dexter, T.M. (1986). *Lymphokine Res.* **5**, 97–104.

Matsuda, S., Shirafuji, N. and Asano, S. (1989). *Blood* **74**, 2343–2348.

Mavilio, F., Kreider, B.L., Valtieri, M., Naso, G., Shirsat, N., Venturelli, D., Reddy, E. P. and Rovera, G. (1989). *Oncogene* **4**, 301–308.

McColl, S.R., Kreis, C., DiPersio, J.F., Borgeat, P. and Naccache, P.H. (1989). *Blood* **73**, 588–591.

McCulloch, E.A. and Till, J.E. (1964). *Radiation Res.* **22**, 383–389.

Mege, J.L., Gomez-Cambronero, J., Molski, T.F., Becker, E.L. and Sha'afi, R.I. (1989). *J. Leucocyte Biol.* **46**, 161–168.

Metcalf, D. (1984). *The Hemopoietic Colony Stimulating Factors*. Elsevier, Amsterdam.

Metcalf, D., Moore, M.A.S., Sheridan, J.W. and Spitzer, G. (1974). *Blood* **43**, 847–859.

Mochizuki, D.Y., Eisenman, J.K., Conlon, P.J., Larsen, A. and Tuchinski, R.J. (1987). *Proc. Natl. Acad. Sci. USA* **84**, 5267–5271.

Moore, M.A.S. (1990). *Cancer* **65**, 836–844.

Morstyn, G., Lieschke, G.L., Sheridan, W., Layton, J. and Cebon, J. (1989). *Trends Pharmacol. Sci.* **10**, 154–159.

Nair, A.P., Diamantis, I.D., Conscience, J.F., Kindler, V., Hofer, P. and Moroni, C. (1989). *Mol. Cell. Biol.* **9**, 1183–1190.

Naoe, T., Doi, S., Yamanaka, K., Naito, K., Nitta, M. and Yamada, K. (1989). *Nippon Ketsueki Gakai Zasshi* **52**, 32–37.

Needleman, S.W., Gutheil, J.C., Kapil, V., Mane, S.M. (1989). *Leukemia* **3**, 827–829.

Nicola, N.A. and Metcalf, D. (1986). *J. Cell Physiol.* **128**, 180–188.

Nienhuis, A.W., Donahue, R.E., Karlsson, S., Clark, S.C., Agricola, B., Antinoff, N., Pierce, J.E., Turner, P., Anderson, W.F. and Nathan, D. G. (1987). *J. Clin. Invest.* **80**, 573–577.

Nowell, P.C. and Hungerford, D.A. (1960). *J. Natl. Cancer Inst.* **25**, 85–93.

Padua, R.A., Carter, G., Hughes, D., Gow, J., Farr, C., Oscier, D., McCormick, F. and Jacobs, A. (1988). *Leukemia* **2**, 503–510.

Ponticelli, A.S., Whitlock, C.A., Rosenberg, N. and Witte, O.N. (1982). *Cell* **29**, 953–960.

Priest, J.R., Robison, L.L., McKenna, R.W., Lindquist, L.L., Warkenter, P.J., LeBien, T.W., Woods, W.G., Kersey, J.H., Coccia, P.F. and Nesbit, M.E. (1980). *Blood* **56**, 15–22.

Ramakrishnan, L. and Rosenberg, N. (1989). *Biochim. Biophys. Acta* **989**, 209–224.

Richter, J., Andersson, J. and Olsson, I. (1989). *J. Immunol.* **142**, 3199–3205.

Risser, R. and Holland, G.D. (1989). *Curr. Topics Microbiol. Immunol.* **147**, 129–153.

Rosenthal, S., Canellos, G.P. and Gralnick, H.R. (1977). *Am. J. Med.* **63**, 116–124.

Saglio, G., Serra, A., Novanno, A., Falda, M. and Gavosto, F. (1989). *Tumori* **75**, 337–340.

Schreurs, J., Arai, K-I. and Miyajima, A. (1990). *Growth Factors* **2**, 221–233.

Sengupta, A., Liu, W-K., Yeung, Y.G., Yeung, D.C.Y., Frackleton, A.R. and Stanley, E.R. (1988). *Proc. Natl. Acad. Sci. USA* **85**, 8062–8066.

Senn, H.P., Tran-Thang, C., Wadmar-Filipowicz, A., Jiricnyl, J., Fopp, M., Gratwohl, A., Signer, E., Weber, W. and Moroni, C. (1988a). *Int. J. Cancer* **41**, 59–64.

Senn, H.P., Jiricnyl, J., Fopp, M., Schmid, L. and Moroni, C. (1988b). *Blood* **72**, 931–935.

Sherr, C.J. (1990). *Blood* **75**, 1–12.

Shtivelman, E., Lifshitz, B., Gale, R.P. and Canaani, E. (1985). *Nature* **315**, 550–554.

Singer, J.W., Arlin, Z.A., Najfeld, V., Adamson, J.W., Kempin, S.J., Clarkson, B.D. and Fialkow, P.J. (1980). *Blood* **56**, 356–360.

Srivastava, A., Boswell, H.S., Heerema, N.A., Nahreini, P., Lauer, R.C., Antony, A.C., Hoffman, R. and Tricot, G.J. (1988). *Cancer Genet. Cytogenet.* **35**, 61–71.

Strife, A., Lambek, C., Wisniewski, D., Wachter, M., Gulati, S.C. and Clarkson, B.D. (1988). *Cancer Res.* **48**, 1035–1041.

Vallance, S.J. and Whetton, A.D. (1991). *Biochem. Soc. Trans.* **19**, 307–315.

Whetton, A.D. and Dexter, T.M. (1989). *Biochim. Biophys. Acta* **989**, 111–132.

Williams, G.T., Smith, C.A., Spooncer, E., Dexter, T.M. and Taylor, D.R. (1990). *Nature* **343**, 76–79.

Wong, G.G., Witek-Giannotti, J.S., Temple, P.A., Kriz, R., Ferenz, C., Hewick, R.M., Clark, S.C., Ikebuchi, K. and Ogawa, M. (1988). *J. Immunol.* **140**, 3040–3044.

Yunis, J.J., Boot, A.J., Mayer, M.G. and Bos, J.L. (1989). *Oncogene* **4**, 609–614.

Index